SECOND EDITION

Updated for ArcGIS Pro 2.3

GETTING TO KNOW

ArcGIS® PRO

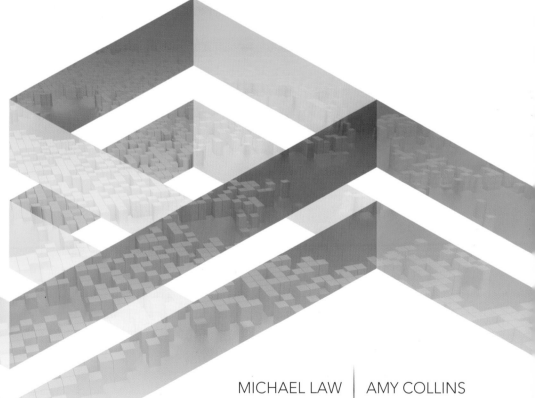

MICHAEL LAW | AMY COLLINS

Esri Press

REDLANDS | CALIFORNIA

Esri Press, 380 New York Street, Redlands, California 92373-8100
Copyright 2019 Esri
All rights reserved. First edition 2016
23 22 21 20 19   1 2 3 4 5 6 7 8 9 10

Printed in the United States of America

Library of Congress Cataloging-in-Publication Data

Names: Law, Michael, 1980- author. | Collins, Amy, 1974- author.
Title: Getting to know ArcGIS pro / Michael Law and Amy Collins.
Description: Second edition. | Redlands, California : Esri Press, [2019] |
    Includes index.
Identifiers: LCCN 2018042818| ISBN 9781589485372 (pbk. : alk. paper) | ISBN
    9781589485389 (ebook)
Subjects: LCSH: ArcGIS. | Geographic information systems. | Graphical user
    interfaces (Computer systems)
Classification: LCC G70.212 .L34 2019 | DDC 910.285/53--dc23 LC record available at
https://urldefense.proofpoint.com/v2/url?u=https-3A__lccn.loc.gov_2018042818&d=DwIFAg&c=n6-
cguzQvX_tUIrZOS_4Og&r=qNU49__SCQN30XC-f38qj8bYYMTIH4VCOt-
Jb8fvjUA&m=jIaaedGCpLWLyxE8ELtvKBydjB__U7AI6ZRugsV1-
rk&s=U3EWrydVdXSX9jMrwoohUab81H9fHBp6g90LfOMYjoU&e=

Ask for Esri Press titles at your local bookstore or order by calling 1-800-447-9778. You can also shop online at www.esri.com/esripress. Outside the United States, contact your local Esri distributor or shop online at eurospanbookstore.com/esri.

Esri Press titles are distributed to the trade by the following:

In North America:
Ingram Publisher Services
Toll-free telephone: 800-648-3104
Toll-free fax: 800-838-1149
E-mail: customerservice@ingrampublisherservices.com

In the United Kingdom, Europe, the Middle East and Africa, Asia, and Australia:
Eurospan Group   Telephone 44(0) 1767 604972
3 Henrietta Street   Fax: 44(0) 1767 6016-40
London WC2E 8LU   E-mail:eurospan@turpin-distribution.com
United Kingdom

Contents

Preface

The title sums it up: this book is for those who want to get to know ArcGIS Pro—a new generation of GIS software from Esri. Whether you are a student in an introductory GIS course, an at-home learner who wants to build a foundational knowledge of GIS, or a professional who is considering adding GIS to your arsenal, this book is for you. No prior GIS software knowledge is required or assumed. This book is also suitable for those who are used to a different GIS product and want to see how to do familiar tasks in a new environment.

The primary focus is, naturally, ArcGIS Pro, but because of the integrated design of the ArcGIS® platform, other ArcGIS components are incorporated as well, such as ArcGIS Online and some mobile apps. ArcGIS Online is integrated into most of the chapters with the use of web-serviced basemaps. In light of this feature, an internet connection is strongly recommended.

A word about scope—although this workbook is designed to provide a broad overview of ArcGIS Pro, a truly comprehensive manual would be massive. Instead, we aim to provide a diverse sampling of industries, scenarios, and workflows that highlight the broad appeal and many core functions offered by GIS and ArcGIS Pro. At the same time, we try to keep the book's length reasonable—something that a student in a classroom can feasibly complete in a quarter or semester. When you complete this book, you should feel comfortable enough with ArcGIS Pro to start working with it on your own.

About the second edition

The second edition has been updated throughout with new screenshots, instructions, and projects to work with the updated software. Some datasets were updated as well to show newer data trends.

Book features

The book has 10 chapters, each containing the following features, which are designed to facilitate an efficient and effective learning process.

Exercise objectives

Exercises are composed of learning objectives, which are listed for each chapter and repeated as headings throughout the chapter exercises. Each objective is accomplished by following a sequence of steps. Using objective headings helps break each exercise into logical chunks and provides a reminder of why you are clicking this or that button, option, or command.

Data list

In the real world, you do not begin a geospatial analysis project before first gathering relevant data. Therefore, we list the student data, with a brief description of what it is and where it comes from, at the start of each chapter.

Exercise workflow

As an expansion of the exercise objectives, each exercise begins with a summary of the workflow, explaining the "what, why, and how" of the upcoming exercise. This description will help you understand the bigger picture, rather than get muddled in a sea of instructions.

A note about exercise scenarios—many of them are based on actual real-world projects; however, the data and workflows are usually simplified for training purposes. These exercises are meant to teach software and data management skills in a realistic setting; they are not meant to be an authoritative guide to geographic problem solving.

GIS in the world

These short sidebars highlight real-world GIS problem solving and offer a link to read more.

Tips and questions

Reminders, shortcuts, or alternative approaches are sprinkled throughout each chapter. Questions keep learners actively involved. (Answers can be found in the book's online resources, esri.com/gtkagpro2.)

Summary

The summary offers a brief recap of what you have learned in each chapter.

Glossary terms

Shown in colored text, glossary terms are listed at the end of each chapter and defined in the glossary at the end of the book.

About trial software and exercise data

The ArcGIS® Desktop suite includes two powerful mapping solutions: ArcMap™ and ArcGIS Pro. You can download ArcGIS Pro by going to the book's online resources (esri.com/gtkagpro2). The trial software offers an Advanced license, with optional extensions. You will need the authorization code found on the inside back cover of this book.

Exercises were written and tested using ArcGIS Pro 2.3 on the Windows 10 operating system. To perform the exercises, you must download the exercise data. You will find instructions for downloading the data in online resources (esri.com/gtkagpro2).

> **TIP** *Because many of the exercises require users to modify the original data, we recommend that you make a copy of each chapter's data folder before you start any exercises.*

How to use this book

Each chapter focuses on a unique project and has its own dataset, so theoretically you can do the chapters in any order. But the book is designed for linear progression—that is, chapter 2 has more explanation and more explicit instruction than chapter 9, in which we assume a more GIS-savvy audience. Also, exercises within chapters typically build on each other, so it is advisable to do all the exercises in a chapter in order. If you cannot complete an exercise successfully, most chapters provide interim data (in a Results folder) so that you can continue with the remaining exercises.

A note about language

When we use the word "click," understand that it can mean either *click* or *tap*, depending on whether you are using a mouse, touchpad, or touchscreen computer. For the sake of brevity, "click" is used instead of "click or tap."

Acknowledgments

It takes a village to make a book. We are indebted to many individuals at Esri for contributing to the process. Thanks to everyone at Esri Press. And thank you to all the reviewers and testers of this book. Also, to all the individuals and organizations who provided data, graphics, project scenarios, and advice: thank you. This book would not be what it is without your assistance and generosity. A complete list of data contributors can be found in appendix A.

And thanks to the GIS learners who purchase this book. We hope you enjoy the second edition of *Getting to Know ArcGIS Pro*.

Michael Law
Toronto, Ontario

Amy Collins
Napa, California

CHAPTER 1
Introducing GIS

Exercise objectives

1a: Explore ArcGIS Online

- Explore a public map
- Sign in and join an organization
- Configure the map symbology
- Configure map pop-up windows
- Save a map

What is a GIS?

Probably the most commonly asked question to those working in the geographic information system (*GIS*) field is also one of the most difficult to answer in just a few brief paragraphs: What is a GIS? A GIS is composed of five interacting parts that include hardware, software, data, procedures, and people. You are likely already familiar with the hardware—computers, smartphones, and tablets. The software consists of applications that help make maps. The data is information in the form of points, lines, and polygons that you see on a *map*. People, users like you, learn how to collect data using mobile devices and then make maps using the software and data on computers. As your knowledge of GIS grows, you will learn more about procedures and workflows to make maps for yourself or your organization. Decision-makers and others in an organization rely on GIS staff to maintain data and create insightful map products.

GIS has many facets. It captures, stores, and manages data. It allows you to visualize, question, analyze, and interpret the data to understand relationships, patterns, and trends. GIS can be used simply for mapping and cartography. You can use it on the web to view maps and collections of data. You can also use it to perform spatial analysis to derive information from multiple data sources. In any capacity, the results from a GIS can influence decisions. Organizations in almost every industry, no matter what size, benefit from GIS and realize its value.

Collecting spatial data—that is, information that represents real-world locations and the shapes of geographic features and the relationships between them—involves using coordinates and a suitable map projection to reference this data to the earth. For example, the distance that separates a conservation area and a neighborhood of a city is an example of a spatial relationship. How is wildlife in the conservation area affected by the increasing pressures of a growing urban setting? The spatial relationship between geographic features allows the comparison of different types of data.

When paired with attribute data—information about spatial data—a GIS becomes a powerful tool. For example, the location of a hospital is considered the spatial data (referenced to the earth). Information about the hospital, such as its name, number of rooms available, emergency rooms, specialization in medical procedures, patient capacity, and number of staff, is all considered attribute data. You can use this attribute information to maintain records in a hospital network. It allows people who have that information to perform spatial analysis—a technique that reveals patterns and trends—to answer the following types of questions: What are the average wait times for emergency visits to the hospital? Does the patient capacity efficiently serve the demographics of a given city area? Do certain medical conditions happen more frequently in the area, and is the hospital equipped to handle them? To answer these questions fully, you must compare the data and attempt to explain the patterns. A children's hospital can integrate spatial analysis with population-based resource planning to plan children's health-care initiatives. This integration can greatly increase the hospital's ability to identify and allocate resources to better meet local health-care needs, providing timely access to care for children across a city or region.

GIS IN THE WORLD: CULTURAL HERITAGE AND GIS

Cultural Site Research and Management and the University of Arkansas Department of Geosciences began a project more than 20 years ago to understand the accumulating effects of nature and foot traffic in the ancient city of Petra, Jordan, a stone city famously used as a backdrop in the movie *Indiana Jones and the Last Crusade*. GIS was used to create a database and model for use in archaeological research and land management and is currently used to automatically monitor and detect changes in the condition of historical sites in Petra. Read more about the project, "Cultural Heritage Management and GIS in Petra, Jordan," at: www.esri.com/news/arcnews/summer12 articles/cultural-heritage-management-and-gis-in-petra-jordan.html.

Ad-Deir (the Monastery) high above the valley—one of the largest hewn structures in Petra.
Photo courtesy of Thomas Paradise.

Oblique view of Petra. Blue lines represent ephemeral watercourses (wadis), and red cubes represent GPS markers at primary tomb facades. The map was created by Christopher C. Angel in ArcGIS (2012).

GIS today

GIS has been helping people better understand their world since the 1960s. It provides a framework of practical means for transforming the world with all kinds of activities, from improving emergency response to understanding bird migration patterns. People are integrating GIS into how they work with data because it is a visual, quantitative, and analytic tool. It provides people with the structures and concepts to handle data systematically.

People today have unprecedented access to data and information. A growing system of connected networks allows people to easily access data, collaborate with others, and produce and share results from desktops, laptops, and mobile devices—essentially from anywhere. The current trend of connecting people who work in the office and in the field allows for real-time analysis. Decision-makers use operations dashboards to monitor real-time data feeds and other sources of information. For example, a GIS coordinator for a local government can track real-time emergencies and respond by coordinating fire, police, and ambulance resources.

GIS is pervasive, interactive, and social. Dynamic and interactive maps on the internet, known simply as *web maps*, are ideal for allowing many users to access and quickly locate features and visualize data. In the past, it took a team of GIS professionals to put together an online map. Now anyone can connect to ArcGIS Online, make a map with a few layers and a *basemap*, and then share it with friends, coworkers, or anyone. The latest generation of web maps has simplified that process and now forms a platform that anyone can use.

Governments are opening access to data at an unprecedented rate. The *open data* movement provides agencies and the public with authoritative data and enables all levels of government to develop new tools and applications. Typically, only highly sensitive data is safeguarded or copyrighted anymore. Open data provides a way for people to extract information when they need it. It allows citizens, organizations, and governments to get right to problem solving, rather than spending a great deal of valuable time searching for and requesting data. ArcGIS® Open Data, an ArcGIS Online solution, allows an organization to host the data it collects so that the public can freely view interactive maps and search for and download data.

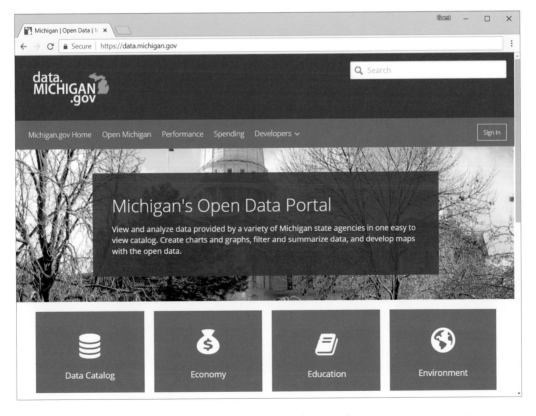

Governments and organizations share open data through online portals. Michigan Open Data Portal (http://data.michigan.gov).

GIS IN THE WORLD: FIRE PREVENTION

The City of Surrey Fire Service in British Columbia, Canada, focuses on fire-reduction education programs to increase prevention and awareness and create safer communities. The agency used GIS to identify areas of the city that met the requirements of being at high risk. GIS allowed the city to prepare a strategy of targeting education to reduce home fires. Read more about the project, "Preventing Home Fires before They Start," at www.esri.com/esri-news/arcnews/winter1314articles/preventing -home-fires-before-they-start.

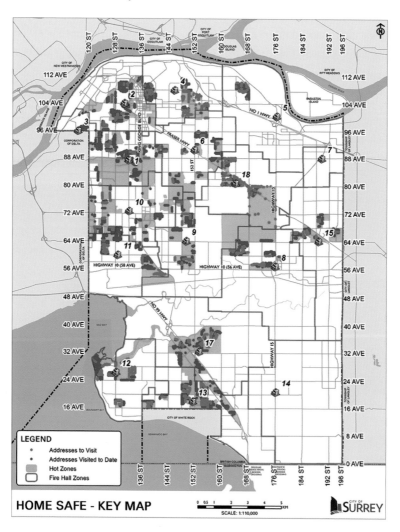

Residential fire incidents in the City of Surrey for a six-year period. Courtesy of the City of Surrey.

Basic GIS principles and concepts

You can visualize data in a GIS as layers in a map. You can represent geographic and manufactured objects on the earth in a map by symbols: points, lines, and polygons. In the accompanying map, points represent trees and points of interest; lines represent roadways; and the polygons represent building footprints, green space, and water. Point, line, and polygon data is also called *vector* data. Features of the same type—such as trees, roadways, or buildings—are grouped together and displayed as *layers* on a map. To make a map, you add as many layers as you need to tell a story. If you are telling a story about a river that seasonally floods, you add a river layer and past flood hazard layers. You can also add a land-use layer to visualize what type of property, such as agricultural or residential, is affected by the flooded river. If you are building a city map, you start with a boundary layer, a street layer, and building footprints. By adding more layers, you can build a map that describes the city to your readers.

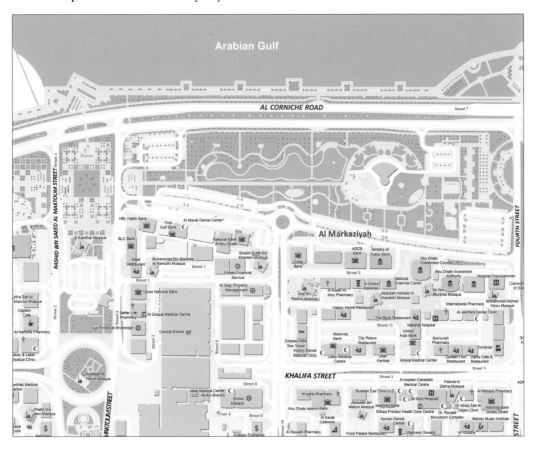

This map of Abu Dhabi shows building footprints, points of interest, roads, green space, and water features. Map courtesy of Municipality of Abu Dhabi City.

If you make a map of your house, a lake, or a city park, you might draw an outline to represent the outer boundary. But what about natural phenomena—such as temperature, elevation, precipitation, ocean currents, and wind speed—that have no real boundaries? Weather maps show blue areas for cold and red areas for hot. You can represent wind speed using a range of colors. You can instead record and collect measured values for any location on the earth's surface to form a digital surface, also known as a *raster*. Captured location data is recorded in a matrix of identically sized square cells at a specific resolution—for example, 15 square meters. In the accompanying example, an analysis of an aquifer uses different rasters to calculate a result showing saturated thickness and usable lifetime.

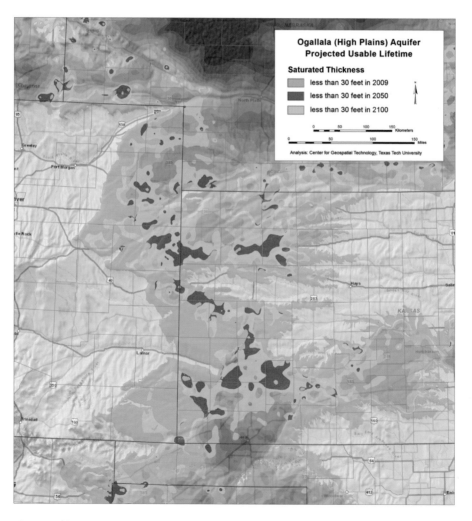

This map of the Ogallala Aquifer shows a surface that represents the saturated thickness, water-level change, and projected usable lifetime of the aquifer. Map courtesy of Center for Geospatial Technology.

Features have locational data behind them. Features also contain attribute data, known as *attributes*. For a forestry map, point features that represent trees might include attributes such as tree species, height, bark thickness, and trunk diameter. For a utility map, lines that represent sewer pipes might include attributes such as flow rate, flow direction, pipe material, and length. Feature attribute information is stored in a table in a GIS database. Each feature occupies a row in the table, and an attribute field occupies a column. A GIS database can hold large collections of features and their corresponding attribute data, also referred to simply as "attributes." A GIS provides many tools for you to query, manipulate, and summarize large quantities of data.

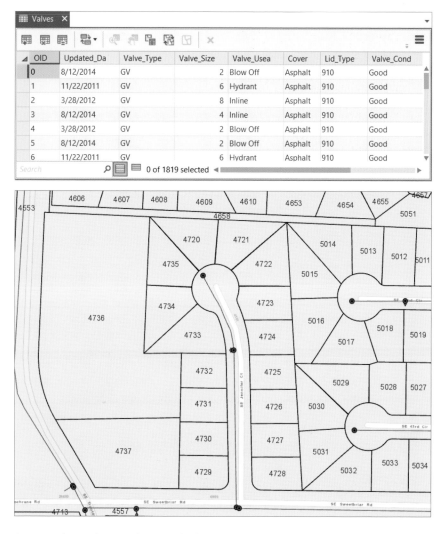

A map with parcels, water lines, and valves. Attributes of the valve features include maintenance dates and information about valve type, size, usage, cover, lid type, and condition.

Data can be queried and analyzed. In a GIS, you can perform a query on all the data that relates to phrases, terms, or features that you choose. For example, you might be looking for clusters of low-income neighborhoods to analyze poverty levels per square mile. Querying data from a database allows you to display only the data that relates to a certain theme. Additionally, a GIS enables you to identify spatial patterns in the data through the use of geospatial processing tools. What is the problem you are trying to solve, and where is it located? The accompanying map shows analysis and a complex pattern of senior citizen out-migration. Depending on your project, you can choose from among hundreds of analysis tools.

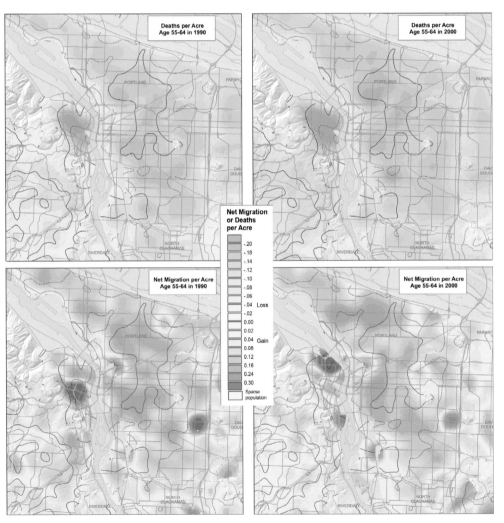

This map of Portland, Oregon, shows net migration or deaths per acre because of "senior shedding" or out-migration. The red isolines identify concentrated areas in which mothers age 30 and over gave birth.
Map courtesy of Portland State University.

The ArcGIS platform

ArcGIS Pro, part of the ArcGIS Desktop suite, is designed for GIS professionals to analyze, visualize, edit, and share maps in both 2D and 3D.

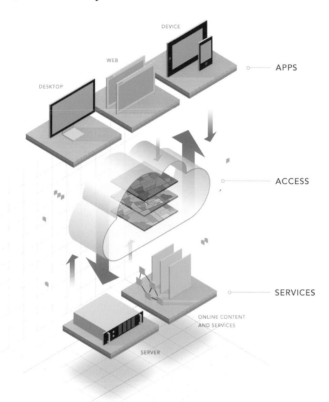

The ArcGIS platform is composed of apps on desktop, web, and mobile devices, which access data controlled by servers and through online content and services.

ArcGIS Desktop is part of the much larger ArcGIS platform, which also includes ArcGIS Online and ArcGIS® Enterprise. Organizations can leverage the entire platform to share maps and apps with their end users.

ArcGIS includes ready-to-use spatial data and related GIS services, such as global basemaps, high-resolution imagery, demographic reports, lifestyle data, geocoding and routing, hosting, and much more.

Finally, the ArcGIS platform includes essential tools for developers to build web, mobile, and desktop apps.

ArcGIS Pro

In this book you will learn how to use ArcGIS Pro and ArcGIS Online. Your work in ArcGIS Pro is organized into *projects*. These projects contain maps, layouts, layers, tables, tasks, tools, and connections to servers, databases, folders, and styles. Essentially, all of the resources needed for a project are in one place. ArcGIS Pro can also connect to ArcGIS Online public content. And if you belong to an organization, you can share the content among your team. Projects are designed to be collaborative so that others can share and open them.

Maps and layouts display a project's spatial data in either 2D maps or 3D scenes, or both simultaneously. You can create, view, and edit multiple maps, layouts, and scenes side by side and even link them so that they can be panned and zoomed together. ArcGIS Pro uses ArcGIS Online basemaps, which provide a backdrop or frame of reference as you add your own layers.

A collection of *geoprocessing* tools allows you to perform spatial analysis and manage GIS data. Geoprocessing involves an operation that manipulates spatial data such as in creating a new dataset or adding a field to a table. You can combine tools in ModelBuilder™ to create a diagram or model of your spatial analysis or data management process. For advanced users, Python, the scripting language of ArcGIS, provides a way to write custom scripting functions to help automate ArcGIS workflows. And tasks can be created and defined for people of an organization who are required to follow specific workflow steps.

The ability to share your work is a central part of the ArcGIS platform. In ArcGIS Pro, you can share maps, layers, or entire projects. Sharing involves packaging components into a compressed file, which you can distribute to others within your organization or externally. You can store your package on a shared network drive, or serve it across a website or mobile device.

To work through the exercises in this book requires that you have ArcGIS Pro loaded on a desktop or laptop computer that is running the Windows operating system, an internet connection, and an up-to-date web browser to access ArcGIS Online. Optionally, you might use a smartphone or tablet to access some of the ArcGIS apps that run on these devices.

EXERCISE 1A

Explore ArcGIS Online

Estimated time to complete: 45 minutes

You will begin your ArcGIS journey by exploring an ArcGIS Online map of geological units in the United States and then configure a new map of bicycle accident data.

Exercise workflow

- *Explore ArcGIS Online to find a public map that shows geological units.*
- *Open a map that shows pedestrian and bicycle accidents.*
- *Configure symbology for accidents and municipal boundaries.*
- *Configure map pop-up windows for readability.*
- *Save the map to your My Content page.*

TIP *Because of the dynamic nature of websites, you should be aware that the appearance or options of ArcGIS Online may change at any time.*

Explore a public map

1. **In a web browser, go to ArcGIS Online at https://www.arcgis.com.**

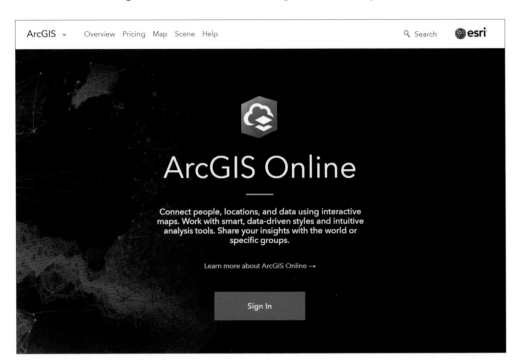

The ArcGIS Online home page is your starting point for learning more about the ArcGIS platform. It has access to information about how ArcGIS Online is used, resources, and help documentation. Even without signing in, you can view and interact with a plethora of community-created public maps and shared data.

ON YOUR OWN

Click the Features tab to learn more about how developers and professionals use ArcGIS Online as well as how you can get maps, apps, and other services.

2. **In the Search box, type** "Geology of United States" **(including the quotation marks), and press Enter or click Go.**

> **TIP** *To be more specific in your search, you can also add the owner as a search parameter. Try typing* **owner:esri_atlas** *after your search term. It will only return maps owned by esri_atlas.*

3. **On the search results page, click the title of the search result named Geology of United States (a web map by esri_atlas).** Clicking the thumbnail opens it in the map viewer.

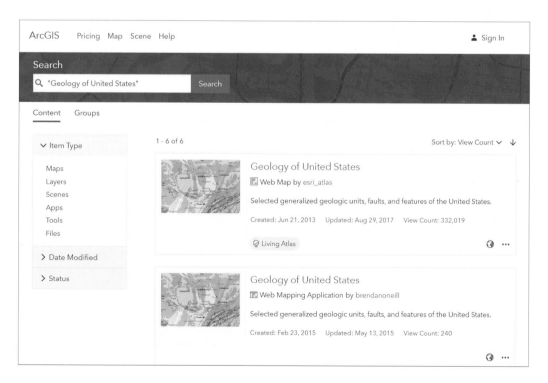

The item details page provides additional information about the map, including a description, layers, and other properties.

4.   Click the Open in Map Viewer button.

The map shows generalized geologic areas and faults for the United States with major geologic features like volcanoes, dikes, and craters. Your map may look different from the one
shown here because Esri and the US Geological Survey maintain the map and keep it current.
The legend shows colors for the geological units.

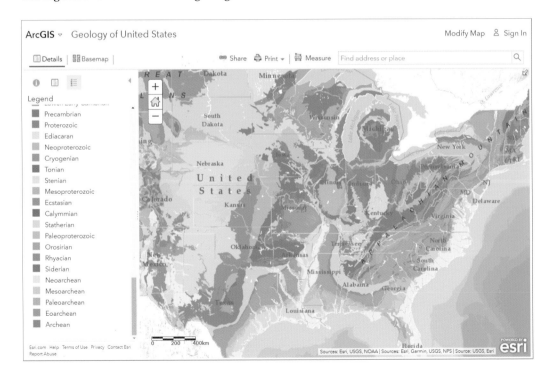

5. **Click a geological unit on the map.**

6. **View the pop-up information.**

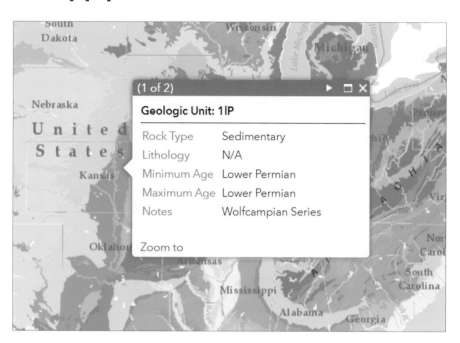

You can see the rock type, age, and associated notes included with the geological unit—all of which are considered attribute data.

7. **Close the pop-up window. Zoom out (with the scroll wheel of your mouse or with the navigation control in the upper-left corner of the map) to see other areas of the United States. Center the map on the state of Alaska.**

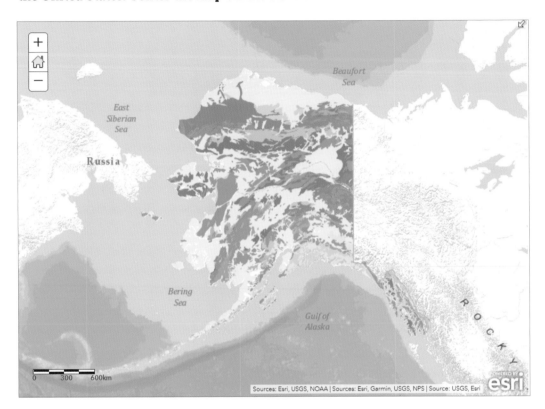

8. **Click Basemap to see available basemaps, and then click Oceans.**

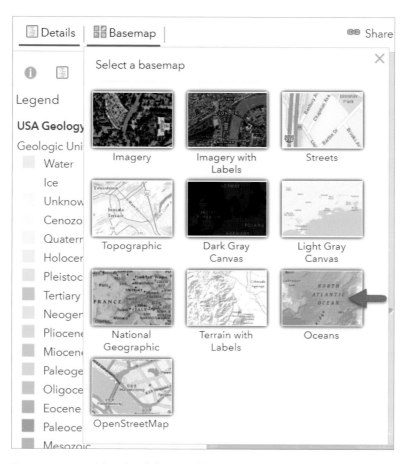

Basemaps provide a backdrop and frame of reference for operational layers such as earthquakes. Now you can see the earthquakes in relation to the ocean floor. As you zoom in, with each zoom level the basemap provides more detail. In this case, you see more details about the ocean floor. In other basemaps, you might see city streets and eventually building outlines. In the satellite basemap, you can see your house, school, work, park, lake, and more.

The side pane to the left of the map contains three tabs: About, Content, and Legend. You can hide the side pane or adjust its width.

9. **Click the Contents button.**

The contents include the USA Geology layer and the Oceans basemap. Layers have a check box so you can turn them on and off. The basemap is always on and has no check box. You can expand the USA Geology layer to see the symbology and other options to show a legend, show a table, change style, and filter data.

10. **In the upper-right corner of the page, click Modify Map.**

You can modify public maps, such as this one, without affecting the author's original map. You can add more layers from ArcGIS Online, the web, or your own computer.

11. **Click Add, and select Search for Layers.**

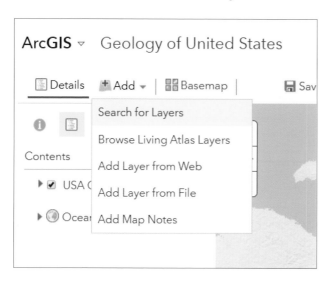

12. **In the Find box, type** World UTM Grid **and press Enter.**

13. **Click the Filter button, and then click Item Type to expand it. Click Feature Layers.**

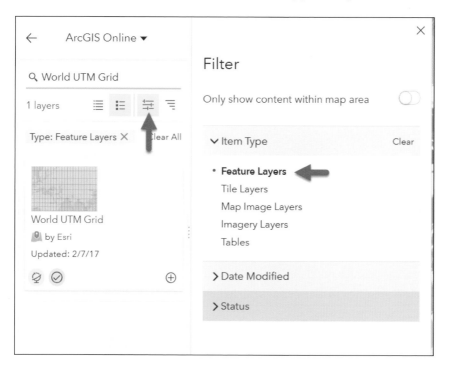

14. **In the search results list, click the World UTM Grid title (the layer should be the only one filtered). In the layer description pane, read the description, and click Add to Map.**

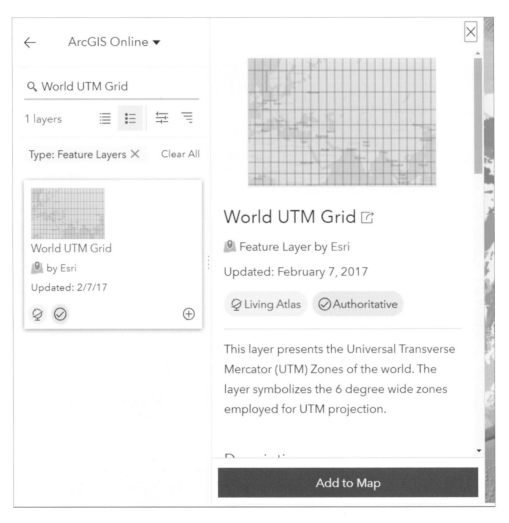

The layer shows Universal Transverse Mercator (UTM) zones of the world, a commonly used projection for many mapping projects because it is a recognized standard around the world. The grid layer provides a quick reference to the six-degree wide zones employed for UTM projection. Because there are so many zones (60 to be exact), you'll use some handy features to find a specific zone more easily.

15. **Click the Back Arrow button to return to the Contents pane. Use the mouse to point to the World UTM Grid layer, and click the Filter button.**

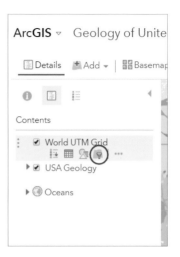

16. **In the filter box, click the first drop-down and choose Zone. In the value field, type** 17. **Click Apply Filter and Zoom To.** Notice how the map zooms out and only shows the filtered result.

17. **Zoom in to a section of the UTM grid layer that overlaps the geological units layer, and click a section of the grid.**

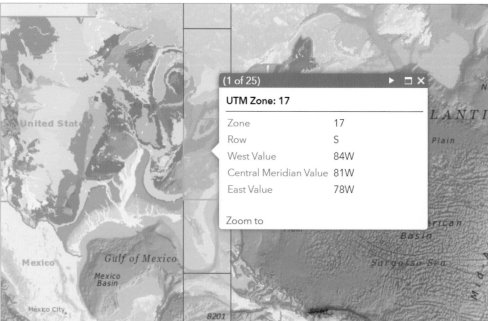

The pop-up shows information about the UTM Zone. In this case, it is Zone 17, which has a western longitude of 84W and an eastern longitude of 78W (the six degrees that make up the zone).

18. **Click the right arrow on the pop-up to see other features. (The number of features may vary from what is shown.)**

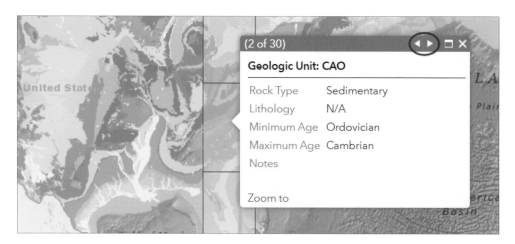

Depending on where you clicked in the grid, other features on other layers may have been detected. Clicking in the grid will help you find features that are within close proximity of each other.

ON YOUR OWN

Look at the USA Geology layer legend and choose a rock type. Create a filter based on that rock type to see where it is most prevalent.

TIP *When you open someone else's map, you cannot save it. You can only share the original owner's map. You can share and copy a shortened link so that you can send the link to colleagues or use the social media links to share on Facebook and Twitter. In later exercises, you will learn how to share your own maps.*

Share

Choose who can view this map.

Your map is currently shared with these people.

- ☐ Everyone (public)
- ☑ Your ArcGIS Organization
- ☐ Members of these groups:

> ☐ Getting to Know: ArcGIS Desktop

Link to this map

http://arcg.is/18L1y   [f] Facebook   [🐦] Twitter

☑ Share current map extent

Embed this map

EMBED IN WEBSITE   CREATE A WEB APP

Note: To embed your map, you must share it with Everyone.

DONE

TIP *A web map can also be embedded on a website. You can customize its size as well as map options and custom location or marker symbol.*

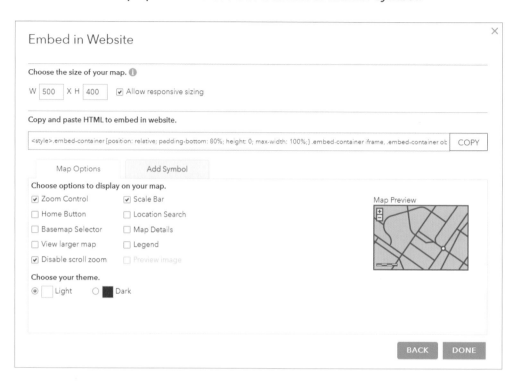

At this point, because this is a public map and you aren't signed into ArcGIS Online, you won't be able to save the map. However, you can still share the map or embed it on a website. In the next section, you will join an ArcGIS Online organization and work with a map as a member, allowing you to save the map as your own.

ON YOUR OWN

Return to the arcgis.com home page and search for other Esri featured content.

Sign in and join an organization

Some exercises in this book require you to sign in to your ArcGIS Online organizational account. To do so, you must first have an ArcGIS Online Public Account. In a web browser, go to https://www.arcgis.com and click Sign In. If you already have an Esri account, you can just sign in. If you do not have one, you can click Create a Public Account and follow the instructions provided. You'll record your account user name and password somewhere safe because you will use them to sign in again throughout this book. To get an organizational account, you'll go to the sign-in page and click Try ArcGIS. An organizational account provides you a suite of ready-to-use apps that run on browsers, desktops, and mobile devices; access to maps and data; 200 ArcGIS Online service credits; and more.

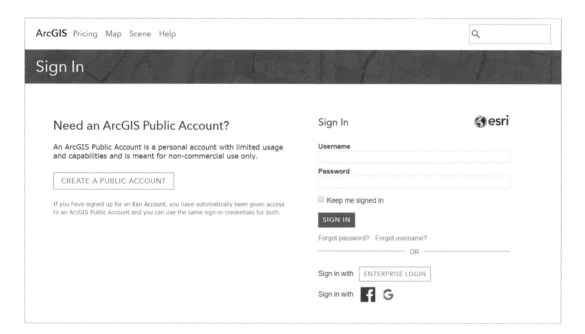

You will visit ArcGIS Online using an organizational account to explore accident data in the City of Wilmington, Delaware.

1. **Go to https://www.arcgis.com.**

2. **Sign In using your ArcGIS Online account credentials. (If necessary, click OK to agree to the conditions to join an organization.)** Notice that you were redirected to a different URL that represents an organization named http://<yourusername>.maps.arcgis. com/home/organization.html. This is your organization's main page, in which you edit settings (such as changing your organization's name), invite other people to join, and see administrative information such as your license and subscription status.

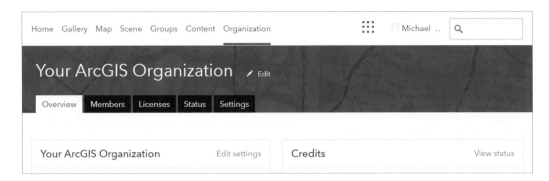

Your trial account automatically makes you the administrator of your organization. As administrator, you have access to everything in your organization, including maps, data, and other resources. You also see a customized view of the site. Organizations have private content, which is available only to their members, or public content, which is available to the entire ArcGIS Online community.

TIP *Your profile contains your user settings. It allows you to store your own content. For more information on how to manage your profile (http://doc.arcgis.com/en/arcgis-online/reference/profile.htm), consult ArcGIS Online Help. Go to http://doc.arcgis.com/en/arcgis-online, and navigate to Reference > Account > Manage profile.*

3. **In the upper-right corner, type** Pedestrian and Bicycle Accidents - Wilmington, DE **in the Search box, and press Enter.**

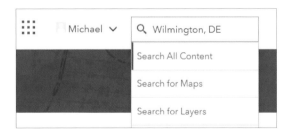

No results are found. This result happens because the search defaults to searching only within your organization. You will disable that option to find other maps that are in ArcGIS Online and available to everyone.

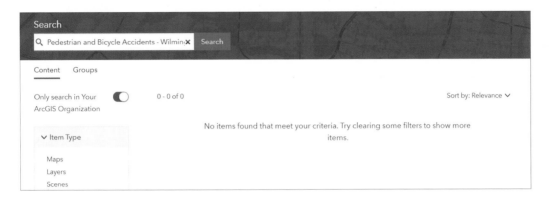

4. **Under Content, turn off the option to Only search in your ArcGIS Organization.**

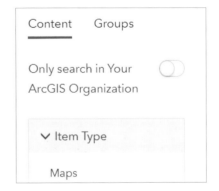

5. **A web map named Pedestrian and Bicycle Accidents - Wilmington, DE will appear in the search results. Click the title to see its description page.**

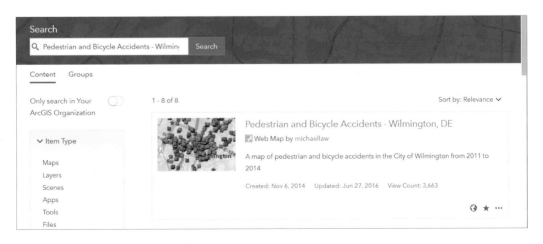

The map description page provides an overview about the map, including the owner, description, map layer contents, and other properties.

6.   **Click Open in Map Viewer (or click the thumbnail image). When the map opens, click the Contents button.** The map shows pedestrian and bicycle accidents from 2011 to 2014 in the City of Wilmington, Delaware. The Contents pane contains six layers—a layer for each year, plus a municipal boundaries layer and a Streets basemap provided by ArcGIS Online. The accident data was provided by a member of the ArcGIS Online community and publicly shared. Open data is often provided by governments, agencies, or individuals for people who want to use it for collaboration, analysis, and research.

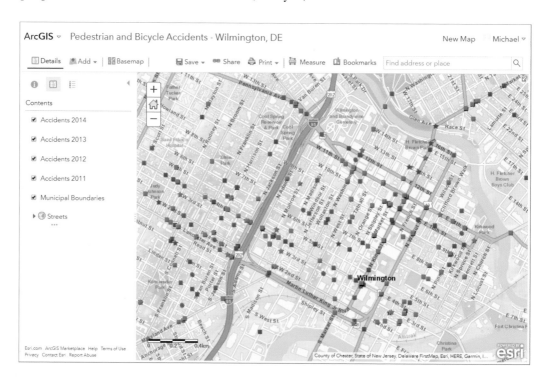

7.   **Move the Accidents 2014 layer to the top of the contents.**

8.   **Hide all layers except the Accidents 2014 layer.** The only points that remain on the map are the blue square and star symbols. Next, you will change the layer symbology to something more meaningful.

Configure the map symbology

1. **To view the current symbology of the Accidents 2014 layer name, click the Show Legend button below the layer name.** A blue-square symbol represents pedestrian accidents, and a blue-star symbol represents bicycle accidents.

2. **Under the Accidents 2014 layer name, click the Change Style button.** A check mark shows that the layer is symbolized based on Types (Unique symbols) to display the Pedestrian attribute, which has only two unique values: pedestrian and bicycle. Next, you will change the symbols to something more meaningful.

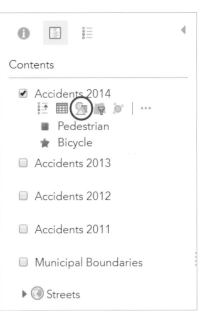

3. **Under Select a drawing style, click Options in Types (Unique symbols).**

4. **Click the Pedestrian blue-square icon.**

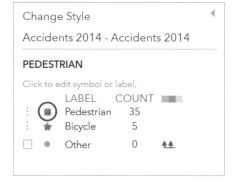

5. **Select Transportation from the drop-down list, and select the red exclamation point in a triangle symbol. Adjust the symbol size to** 18, **and click OK.**

6. **Similarly, choose an appropriate symbol and size for the Bicycle icon (the yellow exclamation point in a triangle symbol is suitable), and click OK.**

7. **Click OK, and then click Done to return to the Contents pane.**

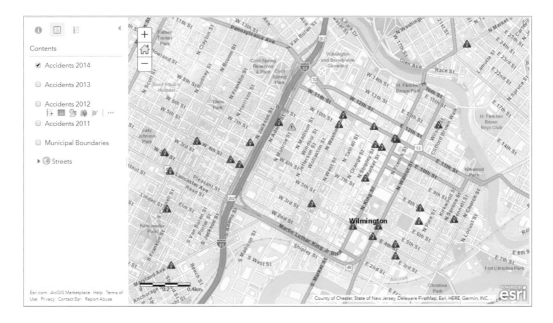

8. **Make the Municipal Boundaries layer visible. Below the Municipal Boundaries layer name, click the Change Style button.**

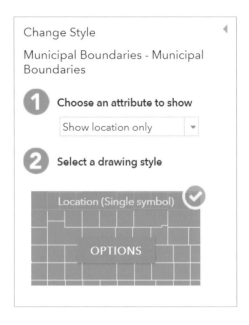

9. **Click Options.**

10. **Click the area symbol.**

11. **Click Fill, and select a light-purple color. Adjust the Transparency slider to about 50%.**

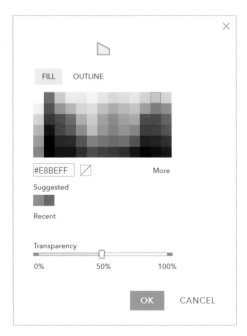

12. **Click Outline, and select a purple color. Adjust the Transparency slider to about 75% and outline width to** 3 **px (pixels), and then click OK.**

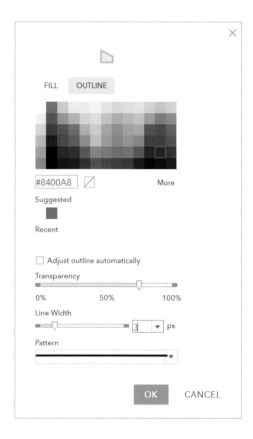

13. **Change the Overall transparency to about 50%.**

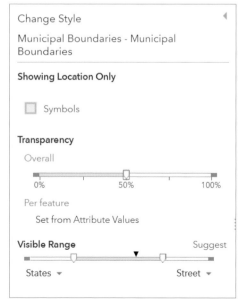

14. Click OK, and then click Done to return to the Contents pane.

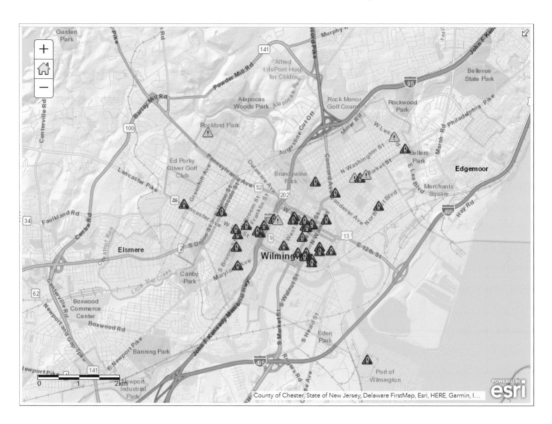

You can now see the municipal boundary more clearly with detail from the basemap still visible.

ON YOUR OWN

Experiment with symbolizing the other layers. Pay attention to how symbol size and map scale relate to each other and how they appear on the map.

You will customize this map further by configuring descriptive information about features in map pop-up windows.

Configure map pop-up windows

1. **Make sure all layers are hidden except Accidents 2014.**

2. **To open a pop-up window for an accident location, click an accident location on the map.**

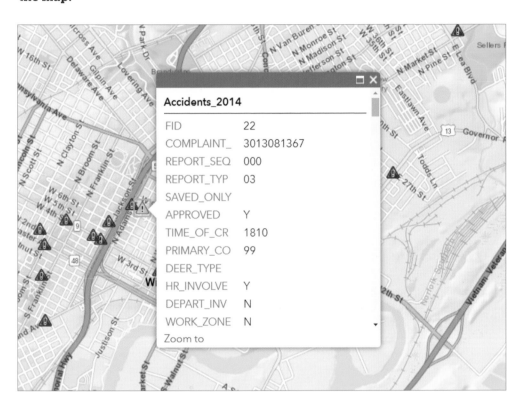

The default pop-up configuration shows an unformatted list of fields and values sourced from the layer's attributes. Although it is useful to know all of this information, for display purposes it is unnecessary to show all of it. You will adjust what attribute fields are displayed so that it is clearer as to what type of accident occurred.

3. **Close the pop-up window. Under the Accidents 2014 layer, click the More Options button and select Configure Pop-up.** The pop-up title is Accidents_2014. You will set a more suitable name for the title instead.

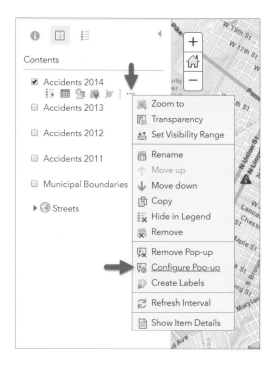

4. **In the Pop-up Title box, type** Accident Details. In the Pop-up Contents section, a list box contains all of the field attributes that are displayed in the pop-up window. It is not necessary to display all of them, and you will configure the pop-up window to display only the information that is relevant.

5. **Under the list box of attributes, click Configure Attributes.** The Configure Attributes window allows you to adjust field visibility and field alias as well as provide a way to move the order of fields.

6. **In the Display column, clear all of the check boxes (click the top display check box twice—to select all and then clear all).**

7. **Scroll down the list and select the Display check box for {LOCATION_4}. Rename the field alias to** LOCATION.

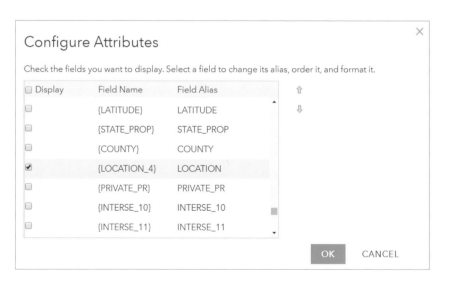

8. Similarly, select the Display check box for {HARMFUL_EV} and rename its field
    alias to INVOLVES. **Do the same for {DATE_OF_CR} and rename it to** DATE.

9. Click OK in the Configure Attributes window and then click Save Pop-up.

10. Click an accident point on the map to see its new pop-up window.

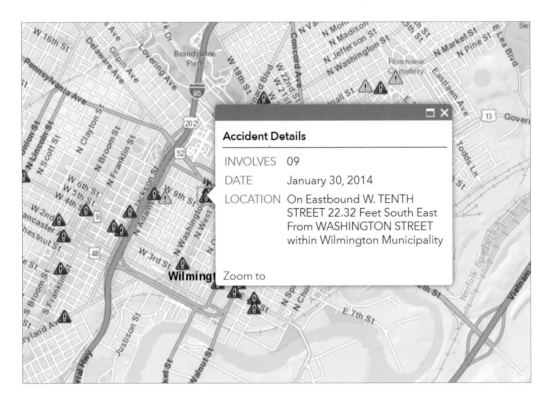

The pop-up window is now much more reader-friendly. In this particular dataset, a code is
used to indicate a pedestrian (09) or a bicycle (10). Pop-up windows can also contain images
and charts.

11. Close the pop-up window.

Save a map

As an organizational account user, you can save the map to your own workspace. By default, maps that you make in the map viewer are private (only you can see them). You will save your map and add it to your own content.

TIP *You are allowed to save a copy of any map that you work on unless the original map author has enabled Save As protection. You cannot update any existing layers that belong to the original map author. However, you can save any changes you make to the map, which then only you can see.*

1. **Click the Save button and then Save As.**

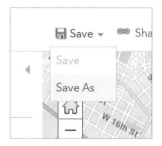

2. **In the Save Map dialog box, change the title to** City of Wilmington Pedestrian and Bicycle Accidents 2011 to 2014.

3. **Add several tags:** pedestrian, accidents, **and** City of Wilmington.

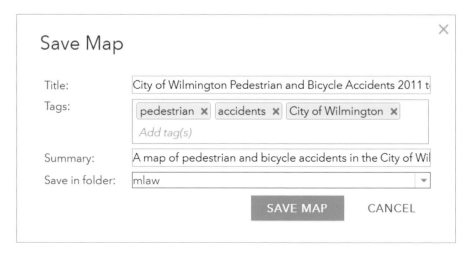

4. **Click Save Map.** You have saved your first ArcGIS Online map. All saved maps appear in My Content and are accessible at any time. Even though the original map was shared with everyone, the map you worked on is considered private and is not shared with anyone in the organization, a group, or the public. You do not need to share this map right now. However, you will learn how to share maps and other content throughout this book as this process is an essential part of the ArcGIS platform.

ON YOUR OWN

Go to My Content and find your own map.

Summary

This chapter introduced you to some background information about what GIS is and how the ArcGIS platform is structured. You started with opening a public map, one that everyone has access to, to see how easy it is to view data and basemaps. You learned how to sign into your organizational account and opened a map that was saved to your account. You learned how to configure symbology and pop-up windows to make the data layers and their attributes more usable. You then learned how to save the map to your own organizational account space.

You should be getting comfortable using ArcGIS Online. You will use it throughout the book, in which you will learn how to share layers, export content to your organizational account, and open maps in ArcGIS Pro. In chapter 2, you will take a first look at ArcGIS Pro.

Glossary terms

GIS
map
basemap
open data
vector
layer
raster
attribute
ArcGIS Pro
project
geoprocessing

CHAPTER 2
A first look at ArcGIS Pro

Exercise objectives

2a: Learn some basics

- Start a new project
- Import a map document
- Create a folder connection
- Modify map contents
- Explore the map
- Examine the contextual ribbon
- Examine feature attributes
- Select features

2b: Go beyond the basics

- Modify feature symbols
- Label features
- Measure distances
- Add a basemap
- Package and share the map

2c: Experience 3D GIS

- Start a new project
- Add data and create a bookmark
- Create a 3D scene

ArcGIS Pro, the latest evolution of the Esri line of GIS software products, is a solution for today's GIS professional. It offers 2D and 3D visualization and analysis within an intuitive, easily navigable interface. ArcGIS Pro seamlessly integrates with networks and the cloud to allow researching, developing, sharing, publishing, and collaborating on GIS projects. The desktop application is designed to be used with ArcGIS Online; for instance, maps authored in ArcGIS Pro can be published to ArcGIS Online, shared with and modified by other users, and then brought back into

ArcGIS Pro. You already got an introduction to ArcGIS Online in the first chapter, so now it is time to meet the main character of this book.

ARCGIS PRO SYSTEM REQUIREMENTS

To take full advantage of the 3D capabilities of ArcGIS Pro, you need an adequate system. Complete system requirements can be found in ArcGIS Pro Online Help, under Get Started > Set up > System requirements, at http://pro.arcgis.com/en/pro-app/get-started/arcgis-pro-system-requirements.htm.

The following short exercises provide an initial overview of ArcGIS Pro. They are designed for brand-new GIS users as well as users who are familiar with other Esri mapping products. You will be introduced to the interface, start exploring some maps, and accomplish some common GIS tasks. These tasks include looking at feature attributes, turning on labels, and modifying map contents. Then, you will work with 3D maps. This chapter will prepare you to successfully complete the GIS project scenarios presented in subsequent chapters.

DATA

You can store your data in a folder called EsriPress. In \\EsriPress\GTKAGPro\World\World.gdb:

- Cities—point features that represent major cities of the world with a population greater than 1 million (Source: ArcWorld)
- Countries—polygon features that represent world countries, including demographic data and air pollution estimates (Source: ArcWorld; US Census International Division, CIA Factbook; The World Bank)
- Latlong—line features that represent latitude and longitude (Source: Esri)
- Ocean—grid of polygons used to display a single-color background behind land features (Source: Esri)

In \\EsriPress\AGPro\3D:

- buildings.shp—polygon building footprints for a section of a Manhattan neighborhood in New York City (Source: New York City Department of City Planning)

EXERCISE 2A

Learn some basics

Estimated time to complete: 40 minutes

In this exercise, you will create a new ArcGIS Pro project, add a map to it, explore map features, and modify the map view.

Exercise workflow

- *Start a new ArcGIS Pro project using an imported map and a new folder connection.*
- *Modify map layers by changing visibility, renaming layers, and moving the order in which they are drawn.*
- *Navigate around the map, and explore information about map features.*
- *Use the Select tool.*

Start a new project

1. **Open ArcGIS Pro (part of the ArcGIS Desktop suite).**

At start-up, you are asked to sign in using your ArcGIS Online organizational account. Your login serves to authorize your ArcGIS Pro license; it also makes it easier to obtain or share data with ArcGIS Online organizations or an ArcGIS portal. You can choose the option to "Sign me in automatically" if you do not want to sign in each time you start the application.

> **TIP** *In the ArcGIS Pro > Licensing settings, you can choose to work offline, in which case a license file serves as authorization rather than your ArcGIS Online credentials. This option is good if you are working without an internet connection. However, keep in mind that you will not see basemaps or any other data layers that are served on the web.*

2. **Enter your user name and password to sign in. (They are the same ones that you registered to use for ArcGIS Online in chapter 1.)**

3. **On the opening screen, choose to open a Map template. Give it a unique name (for example, FirstLook) and save it to either the default or another location, such as your \\EsriPress\GTKAGPro folder. Maintain the default option to create a folder, and click OK.**

> **TIP** *ArcGIS interface elements change periodically to keep up with evolving software functionality, so the images in this book may not exactly match the software interface you see.*

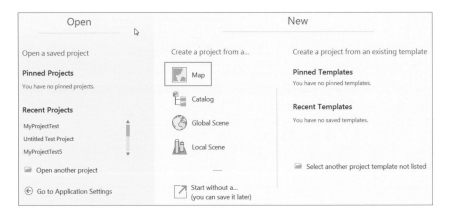

PROJECT TEMPLATES

Using a project template (.aptx) is usually the quickest way to start a project. Templates are shareable project packages, including the specific basemaps, connections, datasets, toolboxes, or add-ins that are most helpful for your project. ArcGIS Pro offers some basic templates (catalog, global scene, local scene, and map), and other industry templates are available as well. You can also create your own template: simply package an existing project and save it to the \Documents\ArcGIS\ProjectTemplates folder in your user profile. You can also share it to your organization's network or through ArcGIS Online. Using an organizational template is a good way to ensure consistency between projects.

TIP *An ArcGIS Pro project is much more than a map document. A project file (.aprx) or project package (.pprx—essentially a zipped-up project file, ideal for sharing) may contain multiple maps, geodatabases, folder connections, layer files, task lists, models, toolboxes, and more. It is like an easily shareable container that holds everything you need for your GIS project.*

You see an empty display area with a ribbon at the top. The ribbon contains tabs that have various buttons, tools, and drop-down options. Notice the elements highlighted in the accompanying graphic. All windows can be resized, docked or detached, or turned off when they are not needed.

Ribbon tabs
(context sensitive)

Tool groups
(context
sensitive)

Contents pane
(docked)

Map display area

Project pane
(floating)

The ArcGIS Pro interface. Tabs and tools change depending on the content you are working with.

To dock or move a pane (such as Contents or Catalog), click the top of the pane, and press and hold the mouse button. As you begin to move the mouse pointer, the pane turns blue, and arrows point to locations in which the pane can be docked. Drag the pane to an arrow, and then release it to choose the docking location. Or drag and release anywhere if you prefer a detached pane.

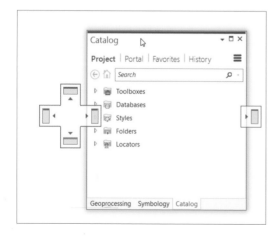

Panes can be stacked to save space—simply click the tabs at the bottom of the panes to switch between them. (By default, the Catalog panes are stacked with the Geoprocessing and Symbology panes; you will explore those tabs later.)

When a pane is docked, click the Auto Hide button ⏚ —it looks like a pushpin—to activate tabbed behavior. A tab replaces the pane when you are working in the map view; just click the tab to reveal the hidden pane.

By default, ArcGIS Pro opens in Catalog view. Since this information is also available in the Catalog pane, you can close Catalog. (Catalog provides additional details and functions you don't need right now.)

4.   Close the Catalog view, and then arrange or resize the panes as you prefer.

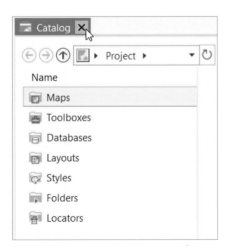

Import a map document

Map documents created with other applications (such as ArcMap) can be imported into an ArcGIS Pro project.

1. **Click the Insert tab, and in the Project group, click the Import Map button.**

    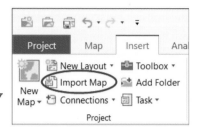

    TIP *If your ArcGIS Pro window is small, you may see only the icon.*

2. **Import World_data.mxd from your \\EsriPress\GTKAGPro\World folder.**

    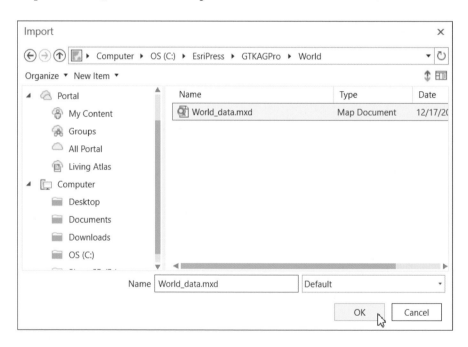

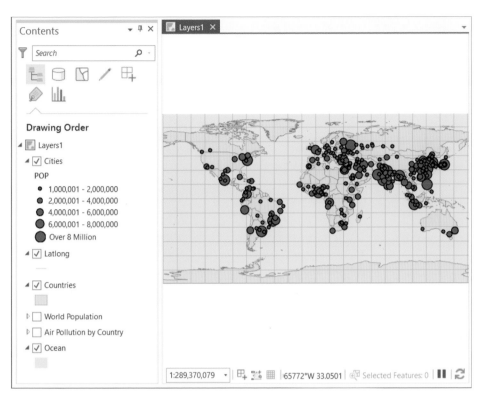

Notice that the Contents pane is populated with the map layers. Some layers are turned on (indicated by a check mark), and others are turned off.

TIP *When the map you import has multiple data frames, each frame becomes a separate map in ArcGIS Pro.*

Create a folder connection

Creating folder connections for your projects allows you to quickly add data or modify a layer's source data. A folder connection prevents you from having to search through multiple system folders.

1. **On the Insert tab, in the Project group, click Add Folder.**

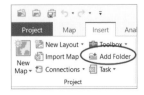

2. **Navigate to the \\EsriPress\GTKAGPro\World folder, and click OK.**

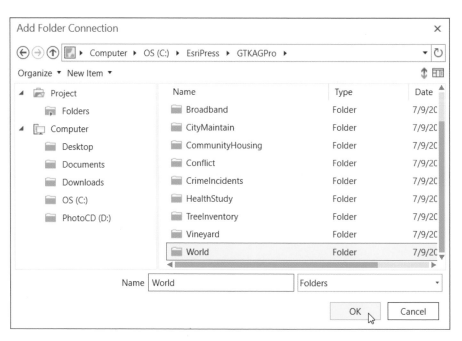

3. **On the Catalog pane, expand Folders to see the new connection.** There is also a connection to the project folder, which is created by default when you start a new project.

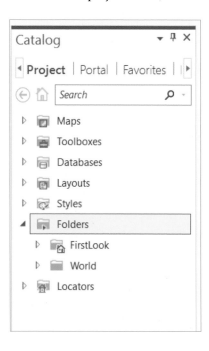

Modify map contents

The Contents pane allows you to modify the map's layers. In the next steps, you will get a quick overview of how to work in the Contents pane.

1. **Look at the Cities layer. Cities are represented by graduated point symbols that correspond to population values—the larger the point, the larger is the city's population. Click the check box to the left of Cities to clear it (thereby turning the layer off in the map).**

2. **In the same way, turn off Countries, and then click the World Population check box to turn it on. Expand the World Population legend by clicking the triangle symbol next to the check box.** The countries are symbolized by population categories using a graduated color ramp—the darker the color, the higher is the population.

   TIP *If you want to remove a layer from the map entirely, and not just turn it off, right-click the layer name and click Remove.*

3. **Collapse all legends.** Collapsing the legends makes it easier to reorder layers in the Contents pane.

4. **Move Latlong below Air Pollution by Country by clicking and dragging to move the layer and release it in the desired location.**

5. **Turn on Air Pollution by Country.** You cannot see this layer because it is underneath World Population. To see the Air Pollution layer, you can turn off World Population or reorder the layers. Think of layers as shapes drawn on sheets of transparent paper. You typically place points and lines above polygons so that they are not covered up.

6. **Turn off World Population to examine the Air Pollution by Country layer. Expand its legend so that you can better understand the map.** This layer shows particulate matter (PM) concentration measurements for each country.

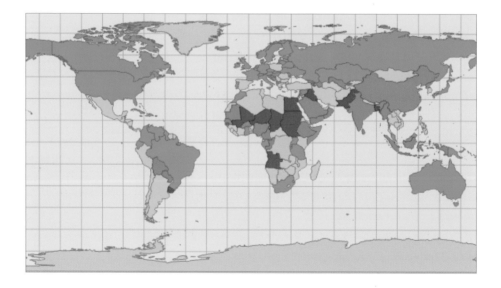

On which continent are PM concentrations highest?

You can find the answers to the questions posed in this book in the book's online resources (esri.com/gtkagpro2).

7. **When you finish, turn on World Population again.** You may also collapse the Air Pollution legend again if you want. Next, you will change the map's name from Layers1 to World.

8. **In the Contents pane, click Layers1 once to high-light it, and click it again to make it editable. Type** World, **and then press Enter.**

Explore the map

To navigate around the map, you will use the Explore tool.

1. **On the Map tab, in the Navigate group, make sure the Explore tool is active.**

> **TIP** *If you want, you can right-click the Explore tool and click Add to Quick Access Toolbar. A small Explore tool icon is added at the top of the application window, so you can easily switch to Explore mode without changing tabs. To undo this change, right-click the icon and click Remove from Quick Access Toolbar. You can add any tool you want to the Quick Access Toolbar.*

2. **Navigate around the map:**
   - **To zoom in and out, move the mouse up and down while pressing and holding the right mouse button. (You can also zoom using the mouse wheel.)**
   - **To pan, press and hold the left mouse button, and drag the map.**

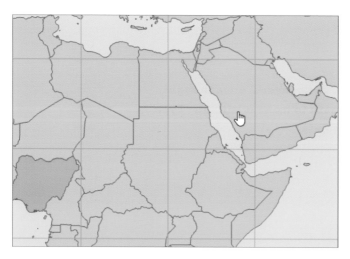

TIP *If you are using a touchscreen monitor, try panning and zooming using conventional touchscreen motions. (Swipe to pan, "pinch" to zoom out, and "reverse pinch" to zoom in.)*

3. **With the Explore tool still active, click any country to see its attributes in a pop-up window. (Zoom and pan as needed.) Notice that the feature first flashes blue and then flashes red.**

As explained in the first chapter, *attributes* are pieces of information about a geographic feature in a GIS. They are usually stored in a table and linked to the feature by a unique identifier. For example, city attributes might include the city name, the county in which it is located, total population, and demographic data. Attribute tables also store information about feature geometry, such as length and area (for lines and polygons).

4. **Close the pop-up window.**

5. **Go to the full extent of the map: on the Map tab, in the Navigate group, click the Full Extent button.**

Examine the contextual ribbon

Before you open an attribute table, you should understand that the ribbon and tools you see in ArcGIS Pro are automatically updated according to context. When you are working with the map view, you have the standard available tabs and tools; but when you work with layers, additional contextual tabs and tools appear.

1. **Make sure the World map frame is selected in the Contents pane.**

The map frame ribbon contains eight tabs.

2. **Click each tab to see the tools and functions offered for each tab. (If you click Project, click the back arrow to return to the map.)**

If you close your Contents pane, how do you restore it?

How do you find a geoprocessing tool?

3. **In the Contents pane, activate and turn on Cities.**

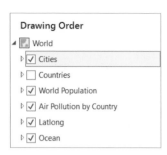

4. **Notice that the Feature Layer ribbon appears, on which there are three additional tabs that relate to feature layers.**

> **TIP** *The Feature Layer ribbon is activated when a layer is highlighted in the Contents pane, regardless of whether the layer is turned on or off.*

Examine feature attributes

1. **Make sure Cities is still active (highlighted) on the Contents pane. On the Feature Layer ribbon, click the Data tab. Notice the functions that are available here.**

2. **In the Table group, click the Attribute Table button.**

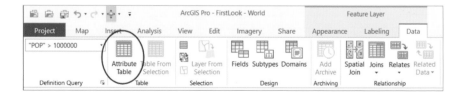

The attribute table opens. It can be docked or floating. The table contains all of the attributes of all of the features in a layer. Compare this table to the Explore pop-up window, which shows only the attributes of a single feature.

TIP *You can also open the attribute table by right-clicking the feature layer in the Contents pane and then clicking Attribute Table.*

3. **Arrange the table as you prefer, and examine the attribute data.**

4. **Right-click the POP heading (scroll as necessary), and click Sort Descending.**

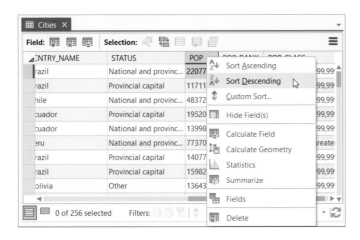

Which city has the largest population?

Select features

1. **Click the gray square to the left of the top row. Notice that the row is highlighted in aqua, and so is the associated feature on the map.** This step is called *selecting* a feature. You can *select* one feature at a time, or multiple features.

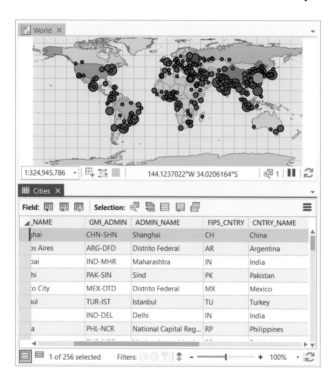

2. **Press and hold the Ctrl key and select the five largest cities in the table.** Notice that the cities are highlighted in the map, and at the bottom of the table you can see how many features are selected. By pressing the buttons shown in the accompanying graphic, you can choose to show all records in the table or only the selected records.

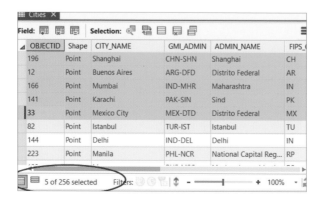

3. **On the Map tab in the Selection group, click the Clear button.** This step
    clears the selection.

4. **Close the table by clicking the X (in the upper-right corner if the table is floating,
    or next to the table name if it is docked).** You can also select features manually in the
    map, without using the attribute table.

5. **Zoom in to any country. On the Map tab, click the Select tool, and click a
    country. Notice that an icon appears. Click the down arrow to see a list of
    overlapping features.**

If you open the World Population attribute table, you see that the selected country's row is
highlighted in the table. Clicking different layers in the overlapping features list changes the
selection and is reflected in the respective feature's attribute table.

TIP *To select multiple features, press and hold the Shift key as you click each
desired feature. Or draw a rectangle (or another shape—click the down
arrow to reconfigure the tool's default behavior) to select all features within
the shape's extent.*

Selections are more than a visual aid; you can generate calculations or run geoprocessing
operations on a selected set of features, rather than the entire set. You will do this later in the
book.

6. **To clear the selection, click the Clear button on the Map tab.**

7. **Zoom to the full extent of the map. At the top of the application, click the Save** **button.** This exercise was meant to get you comfortable navigating the ArcGIS Pro interface. It also underscores the relationship between geographic features and their attributes.

You will continue working with the same project in the next exercise. You can continue to the next exercise, or close ArcGIS Pro and come back to it later.

GIS IN THE WORLD: SKI PATROL USES ARCGIS® TRACKING ANALYST FOR RESCUE MISSIONS

Colorado's Winter Park Ski Resort has successfully used GIS and GPS technology to track lost skiers and snowboarders. Lost resort visitors can be tracked using their GPS-enabled smartphones. GPS points are imported into ArcGIS software, and then the rescue team's locations are tracked using the ArcGIS Tracking Analyst extension so that the mission can be monitored in real time. Read more here, "GIS to the Rescue": www.esri.com/esri-news/arcwatch/0114/gis-to-the-rescue.

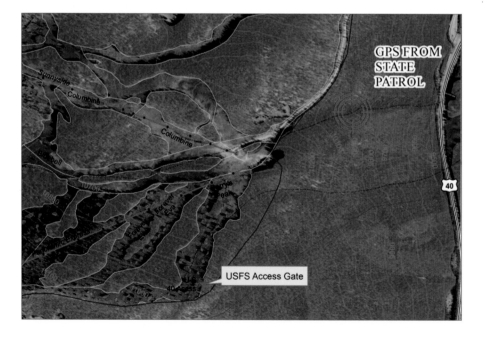

EXERCISE 2B

Go beyond the basics

Estimated time to complete: 20 minutes

In this exercise, you will change feature symbols, configure and display feature labels, use the Measure tool, add a cloud-hosted basemap, and package your project to share online.

Exercise workflow

- *Customize the appearance of the map using symbols and labels.*
- *Use the Measure tool to find the approximate distance between cities.*
- *Examine and add a basemap.*
- *Create a project package for sharing.*

Modify feature symbols

1. **If necessary, open the ArcGIS Pro project you created in exercise 2a.**

2. **In the Contents pane, highlight Cities. On the ribbon, click the Appearance tab, and then click the Symbology button.**

TIP *Here are some other ways to access the Symbology pane:*
- *Right-click a layer in the Contents pane, and click Symbology.*
- *In the layer's legend, click a symbol icon.*
- *If the Symbology tab is already visible (stacked with the Catalog pane, for instance), simply highlight the layer in the Contents pane, and then click the tab.*

The Symbology pane opens. *Symbology* refers to the way GIS features are displayed on a map. The symbology dictates not only the size, shape, and color of map features, but also the conventions and rules used for displaying these features. Symbology is not just for looks; it also conveys meaning to map readers. You have already noticed, for instance, that the Cities layer is represented using *graduated symbols*. In this case, cities are represented by a range of symbols based on an attribute field named POP (which stands for *population*). So, the greater the population, the larger is the symbol.

FEATURE SYMBOL OPTIONS

Spatial data, in its simplest form, is a collection of points, lines, and polygons. But you can modify default symbols to make your map more readable and informative. You can display map features in the following ways:

- Single symbol: one symbol is used for all features in a layer.
- Unique values: used for categorical data, different symbols represent various attributes. For example, imagine you have a polygon layer of parklands, and it has an attribute that identifies each feature as either a state or national park. You can symbolize state parks using a black outline and green fill, and national parks using a gray outline and brown fill.
- Graduated colors: used for quantitative data, different colors represent different value ranges. Often, a color ramp is chosen in which darker colors correspond to higher values. For example, in this map the World Population layer is symbolized using graduated colors, in which darker colors represent higher population attribute values. The user decides how values are classified and how many categories there should be.
- Graduated symbols: used for quantitative data, symbols increase in size with increased values. Earthquake epicenters might be given graduated symbols that represent magnitude values. The user chooses the classification method and number of categories.

- Dictionary: a method to apply symbols—either graphic or text symbols—to data using a predefined data dictionary. Symbols vary according to dictionary rules and attribute values.

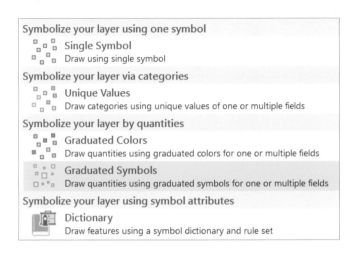

3. **On the Symbology pane, click the symbol template to edit it.**

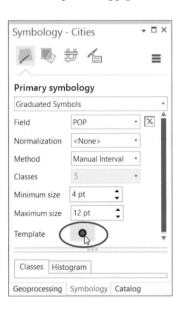

4. **Under Format Point Symbol, choose any 2D symbol template you like from the Gallery, and then click the back arrow to return to the main Symbology pane.**

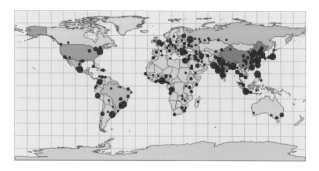

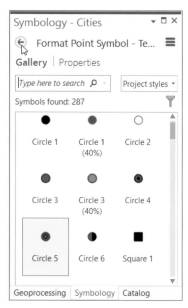

5. **Close the Symbology pane (click the X in the upper-right corner).**

Label features

You will add labels to your map to make it more informative.

1. **In the Contents pane, right-click Cities. In the context menu, click Label.**

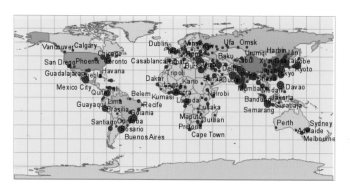

City labels appear cluttered at the full extent of the map. You will set a visibility scale so that city labels appear only when zoomed in.

2. **With Cities still highlighted in the Contents pane, click the Labeling tab in the Feature Layer ribbon.**

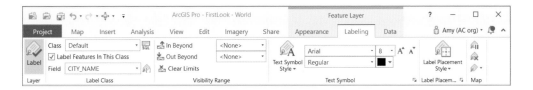

Notice that the labeling field is set to the CITY_NAME attribute. You can change this field so that the labels display city population or some other attribute if you want.

3. **In the Visibility Range group, next to Out Beyond, type** 1:75,000,000, **and then click Enter on the keyboard.** This range means that when you are zoomed to full extent, you do not see the labels, but as you zoom in to a scale of 1:75,000,000 or closer, labels are turned on.

TIP *You can also choose an extent from the option menu.*

4. **Zoom in until the labels appear.**

ON YOUR OWN

Turn on labels for the World Population layer. Modify the default visibility scale if you want.

Measure distances

What if you want to know the approximate distance between Lima, Peru, and Rio de Janeiro, Brazil?

1. **Zoom to South America. (Ensure that you can see Lima and Rio de Janeiro.) On the Map tab, click the Measure tool's down arrow. Notice the different configurations that you can choose for this tool. Click Measure Distance.**

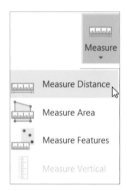

2. **In the Measure tool window, click Options > Distance Units. Click Miles.**

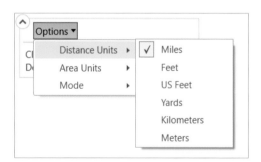

3. **Click the point that represents the city of Lima, and then double-click the point that represents Rio de Janeiro.**

Notice that the tool provides the current segment as well as the entire path length if you click multiple points.

4. **Click the Clear Results button at the top of the Measure window to clear the previous measurement and start again.**

ON YOUR OWN

Measure distances between other cities on the map.

The Measure window remains open as long as the Measure tool is active. Choosing any other tool closes it.

5. **Click the Explore tool.**

Add a basemap

Esri provides several basemaps that you can use in any project. Basemaps are like dynamic graphics of paper maps or imagery. They do not have attributes, but they provide a professional appearance to your map and visual context for your study area. They cover the entire world, so no matter where or at what scale your project is, you can use a basemap. Depending on your project, you might choose to use satellite imagery or a topographic or street map. The basemaps provided by Esri were created collaboratively by several organizations. An organization might create a custom basemap that is used for all of its projects. Or you may decide to not use a basemap at all.

1. **Zoom to the full extent of the map.**

REMIND ME HOW

Click the Full Extent button.

2. **Turn off the Latlong and Ocean layers.**

3. **On the Map tab, in the Layer group, click the Basemap button. From the basemap gallery, select Oceans.**

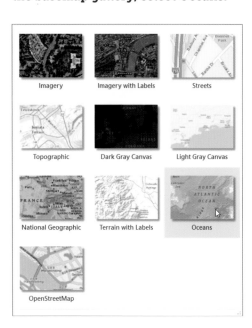

Depending on the size of your window, you may need to zoom in slightly to see the new basemap. Basemaps are hosted by ArcGIS Online. You will not be able to load one if you do not have a good internet connection.

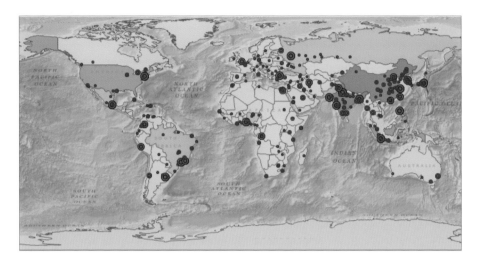

TIP When you zoom in, you may notice duplicate city labels. To avoid double labeling, you can turn off either your Cities labels or Reference-Oceans, a reference layer that comes with the Oceans basemap. Turning off this layer does not affect the basemap, only the labels and political boundaries. To see this reference layer, click the List by Data Source button at the top of the Contents pane.

4. **To remove the old Ocean background layer, in the Contents pane right-click Ocean, and click Remove.**

Package and share the map

1. **Zoom to the full extent, and then save the project by clicking the Save button in the upper-left corner of the window.**

2. **On the Share tab, in the Package group, click the Map button.**

TIP *If you have other project elements you want to share—such as styles, toolboxes, task lists, or attachments—sharing a project package, which includes these other elements, is more appropriate. (You will learn more about styles, toolboxes, and task lists in subsequent chapters.) In this case, sharing a map package is the best choice.*

3. **In the Package Map pane, maintain the default parameters, and then click Analyze. When Analysis is complete, click Package.**

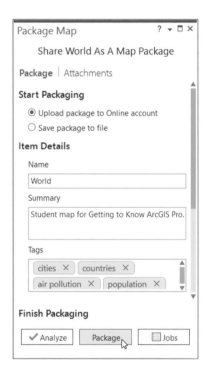

TIP *To package a map, there must be a description in the map's metadata. (To access the metadata, in the Contents pane, right-click the map title, click Properties, and go to the Metadata tab.) There must also be a summary and tags (keywords for searching), which are read from the metadata; if the summary and tags are not in the metadata, you may enter them in the Package Map pane.*

ON YOUR OWN

Go to https://www.arcgis.com and log in to your account. Find the map package under My Content. Click the map, and share it with any group or individual you like.

If you are not continuing to the next exercise, you can close ArcGIS Pro.

EXERCISE 2C

Experience 3D GIS

Estimated time to complete: 20 minutes

Presenting a 3D map engages your audience and provides a wow factor that your 2D maps may lack. But it is not just about looks—with the power of 3D GIS, you can visualize and modify skylines, experience realistic topography, and perform 3D analysis. For example, you can construct sightlines in hilly terrain. In this exercise, you will learn how to convert a 2D map to 3D.

Exercise workflow

- *Start a new project, and add a layer of polygons that represents buildings in New York City.*
- *Use the Bookmark tool.*
- *Convert a 2D map to a 3D map, and then extrude features based on the building height attribute to visualize buildings in a more realistic perspective.*

Start a new project

First, you will start a Local Scene project.

1. **In a new ArcGIS Pro session, choose a Local Scene template; or from an existing project, click the New button at the top of the application.**

2. **Name the project** 3D, **and store it in the \\EsriPress\GTKAGPro\3D folder. Clear the option to create a folder, because a folder is provided for you.**

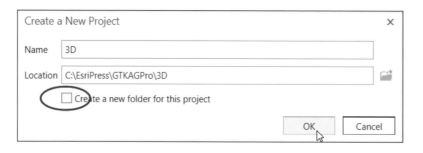

3. **In the 3D project, on the Insert tab, click the New Map down arrow, and click New Map.**

TIP *ArcGIS Pro modifies the New Map default icon to reflect your previous choice, so sometimes you will click only the icon.*

Add data and create a bookmark

Next, you will add a shapefile of some buildings in New York City.

1. **In the Catalog pane, expand Folders, and then expand 3D.** The project automatically stores a folder connection to the location in which you saved your project.

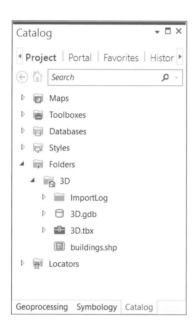

You may also see an Import Log folder in your project workspace.

TIP *To create additional folder connections, right-click Folders and click Add Folder Connection.*

Notice that the 3D folder contains a geodatabase and a toolbox, both of which are currently empty and were created automatically when you created the project. (You will learn more about what these things are in subsequent chapters.) The folder also contains a GIS data file named buildings.shp, which was already there.

2. **In the Catalog pane, click to highlight buildings.shp, and then drag and release it in the map view.**

Building footprints that span 10 blocks of New York City's Upper West Side neighborhood are added to the map, and the map display zooms to the extent of the new layer. Colors will vary, because colors are applied randomly when data is added to a map.

TIP *When data resides in the project folder, it may be the easiest place from which to drag data to the map, as you did here. If the data resides elsewhere, either create a new folder connection or use the Add Data button, navigate to the desired dataset, and click Select. Notice that you can configure the Add Data button to add different types of data (for example, route events or address layers) by clicking the down arrow.*

Next, you will create a bookmark of the buildings layer extent.

3. **On the Map tab, in the Navigate group, click the Bookmarks button, and click New Bookmark. Name it** UWS Buildings, **enter a brief description of your choosing, and then click OK.**

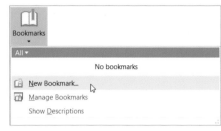

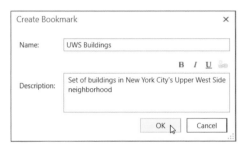

Now when you click Bookmarks, you see a thumbnail along with the name of the bookmark.

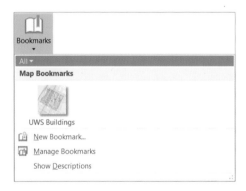

Create a 3D scene

Now you will convert a 2D map into a 3D scene (without losing your 2D map) so that you can visualize the heights of the buildings in the map. ArcGIS Pro basemaps can be used for both 2D and 3D maps. You can turn your buildings layer into a 3D layer because it includes a roof height attribute, measured in feet.

1. **In the Contents pane, right-click Buildings, and click Attribute Table. Scroll to the last column, and sort the Height attribute in descending order.**

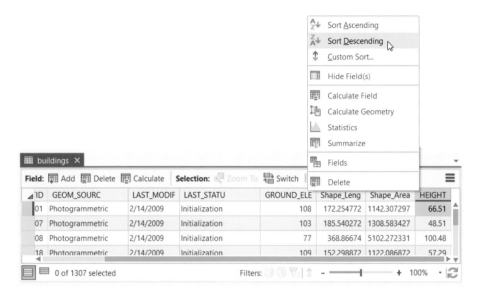

What is the height of the tallest building in this layer?

2. **Close the attribute table.**

3. **On the View tab, click the Convert down arrow, and choose To Local Scene.**

When the conversion is complete, a new 3D scene is added to the project.

Now you will enable 3D rendering for buildings.

4. **In the Contents pane, drag buildings to 3D Layers, and then release.**

5. **In the Contents pane, right-click the buildings layer, and click Zoom To Layer.** Do not worry if the buildings do not render correctly in the 3D map—they must be extruded before they can be displayed properly. Extrusion is the process of stretching flat 2D features vertically so that they appear three-dimensional. You will *extrude* the buildings based on the building height attribute field.

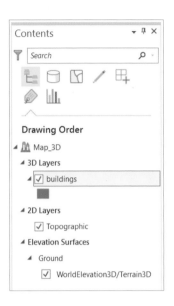

6. **With buildings selected in the Contents pane, on the Feature Layer ribbon click the Appearance tab. In the Extrusion group, click the Type down arrow, and click Base Height. For Field, choose HEIGHT, and for Unit, choose Feet.**

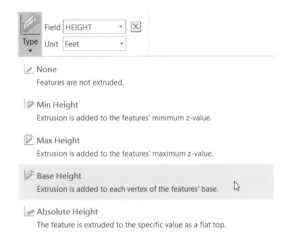

> **TIP** *You are not limited to extruding height measurements. You can create a 3D population map by extruding a population value. Any numeric value can be displayed in 3D.*

TIP *To learn about the different extrusion types, go to the ArcGIS Pro Help topic: Maps and scenes > Layers > Layer properties > Extrude features to 3D symbology.*

7. **On the Appearance tab, click the Symbology tool (or just click the Symbology tab if it's available). In the Symbology pane symbol gallery, click the current symbol, and then scroll down to the ArcGIS 3D style set and select Concrete.**

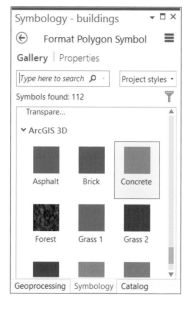

8. **Click the Explore tool (on the Map tab of the main ribbon, or on the Quick Access bar if you placed it there). Point to the Explore tool and familiarize yourself with navigation methods.** You can depress the scroll wheel on the mouse to tilt and rotate a 3D scene.

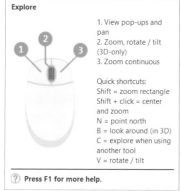

9. **Navigate the 3D scene. Use the mouse wheel to experiment with zooming, tilting, and rotating the scene.** What if you want to present a 2D map alongside a 3D map? You will do that next.

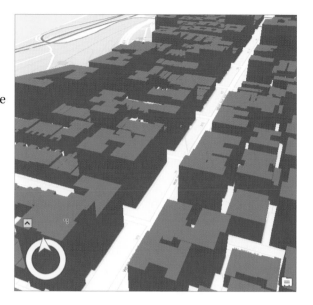

10. **At the top of the map, click the Map_3D tab and drag to place it alongside or below the 2D map.**

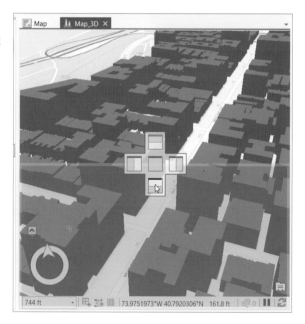

11. **Switch to the View tab, click the Link Views down arrow, and then click Center and Scale.**

Now when you navigate in one map, your view will adjust accordingly in the other map—the views are linked. You can go ahead and test it out.

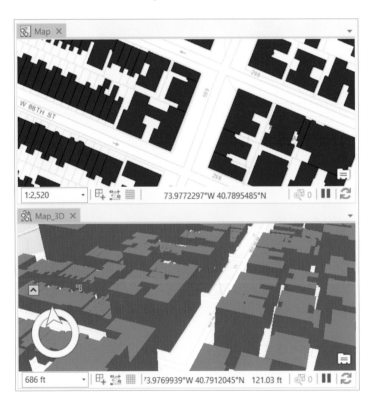

12. **Save and close the project.** If you are not continuing to chapter 3, it is okay to close ArcGIS Pro.

Summary

These exercises gave you a first look at the new ArcGIS Pro application. You learned how to import existing ArcMap map documents, perform some common GIS tasks, add a basemap and local data, and share a project. You also practiced converting 2D data to 3D.

Hopefully, at this point, you are beginning to feel comfortable working with ArcGIS Pro, because it is time to move on. Coming up: integrating desktop and cloud GIS, data investigation, using symbology to reveal data patterns, creating data from scratch, analyzing time and surface variations, deriving new information from existing data, and map presentation tips. As you work through real-world scenarios ranging across industries (for example, health care, conservation, city planning and maintenance, vineyard management, and more), you will develop a better understanding of what ArcGIS Pro can do. You will learn how you can use ArcGIS Pro to streamline operations, enhance workflows, reveal new information, and solve problems in your own community or workplace.

Glossary terms

select
symbology
graduated symbols
extrude

CHAPTER 3
Exploring geospatial relationships

Exercise objectives

3a: Extract part of a dataset

- Add data to a project
- Select features by attributes
- Export the selection to a new dataset

3b: Incorporate tabular data

- Join data tables
- Apply informative symbols
- Import layer symbology
- Use the Swipe function to compare layers
- Overlay additional data

3c: Calculate data statistics

- Add a new field
- Calculate field values
- Display a new field
- Calculate summary statistics
- Examine infographics

3d: Connect spatial datasets

- Relate tables
- Spatially join data

The power of GIS extends far beyond exploring digital maps. You can combine datasets, enrich them with new attributes, derive statistics from them, and obtain new information based on their relationships. In this chapter, you will begin taking advantage of some of the more sophisticated capabilities offered by ArcGIS Pro.

Scenario: you have been hired by a state health coalition that is focusing its efforts on raising awareness about and lowering obesity rates in the state of Illinois. Obesity prevalence has risen dramatically throughout the state over the past decade, along with the incidence of type 2 diabetes, but the prevalence of obesity is higher in some areas than in others. You will explore obesity prevalence rates by county, create visual aids for displaying a year-over-year rising trend, and begin to explore possible reasons why some counties may have higher prevalence rates than others. For instance, is there a link between average income level and obesity levels? What about comparing obesity rates to data that shows limited access to grocery stores—are these "food deserts" more likely to have higher obesity rates?

Your job is to combine obesity statistics provided by the Centers for Disease Control and Prevention (CDC) with county GIS data to do an initial visual analysis of the data, which will eventually lead to more rigorous analysis techniques. The end objective of the analysis is to create informative products that will educate and guide advocates and policy-makers.

GIS IN THE WORLD: AN INTERACTIVE ATLAS OF HEART DISEASE AND STROKE INCIDENCE

The Interactive Atlas of Heart Disease and Stroke (https://nccd.cdc.gov/DHDSPAtlas), created by the CDC and powered by GIS, creates a visual guide for anyone who wants to see where cardiovascular disease occurs, and offers insight about who is at highest risk. Read more about the project here, "CDC Aims at Reducing Annual Heart Disease- and Stroke-Related Deaths in the US by 200,000": www.esri.com/esri-news/arcnews/spring14articles/cdc-aims-at-reducing-annual-heart-disease-and-stroke-related-deaths-in-the-us.

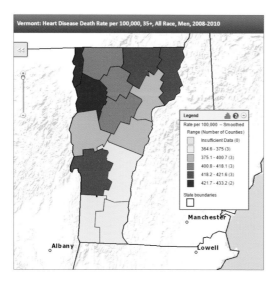

The CDC atlas allows you to apply various filters to get the information you need. Here, the map shows heart disease death rates in Vermont between 2008 and 2010, limited to men above the age of 35 years. This map was created using the Interactive Atlas of Heart Disease and Stroke, a website developed by the Centers for Disease Control and Prevention, Division for Heart Disease and Stroke Prevention. http://nccd.cdc.gov/DHDSPAtlas.

DATA

In the \\EsriPress\GTKAGPro\HealthStudy\Data folder:

- IL_food_deserts.shp—a shapefile that contains polygons that represent low-income areas with limited access to grocery stores *(Sources: Data and Maps for ArcGIS; USDA)*
- IL_med_income.shp—a shapefile that represents Illinois counties with an attribute for median income values *(Sources: Data and Maps for ArcGIS; US Census Bureau)*
- Obesity_Prevalence.dbf—a spreadsheet that contains obesity prevalence rates for Illinois counties from 2004 to 2010 *(Source: Centers for Disease Control and Prevention)*
- us_cnty_enc.shp—a shapefile that contains polygons that represent the county boundaries for the East North Central region of the midwestern United States *(Source: Data and Maps for ArcGIS)*

EXERCISE 3A

Extract part of a dataset

Estimated time to complete: 20 minutes

In this exercise, you set up your project. You will add data to a new project and then do some basic processing to get your data in shape. Once you have completed these steps, you will be ready for a deeper exploration of your data.

Exercise workflow

- *Open a project and add a shapefile of polygons that represent counties from five states in the East North Central region of the Midwest.*
- *Select the features with a STATE_NAME attribute of Illinois.*
- *Export the selected features to a new dataset that contains counties in Illinois only.*

Add data to a project

1. **Start ArcGIS Pro, find the \EsriPress\GTKAGPro\ HealthStudy folder, and open HealthStudy.aprx.** The project template opens with a 2D map that includes the Esri topographic basemap centered on North America.

> **TIP** *A reminder that ArcGIS interface elements change periodically to keep up with evolving software functionality, so the images in this book may not exactly match the software interface you see.*

2. **On the Map tab, ensure that the Explore tool is active, and click any area in North America.** Notice that there is no pop-up window. That is because basemaps function the same way as images—they do not have individual features with attributes.

3. **In the Catalog pane, expand the \HealthStudy\Data folder.**
Notice that this folder contains three shapefiles (.shp), two layer
files (.lyrx), and one database file (.dbf) in addition to two subfold-
ers named Data and Results.

4. **Press and hold Ctrl and click us_cnty_enc.shp and Obesity_ Prevalence.dbf, and
then drag them to the map display.** Your map project now includes a shapefile that contains county features for the East North Central region of the midwestern United States. The area
includes five states—Wisconsin, Michigan, Illinois, Indiana, and Ohio. The project also includes
a nonspatial .dbf table, listed in the Contents pane under Standalone Tables. Because the table
is tabular and does not have spatial attributes, it cannot be displayed on the map in this form.

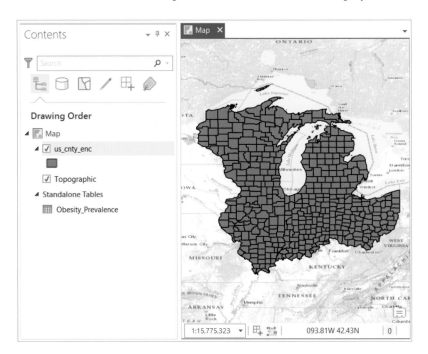

TIP *The colors will vary as they are randomly assigned.*

5. **With the us_cnty_enc.shp layer selected in the Contents pane, click the Data tab, and then click the Attribute Table button.** Examine the table.

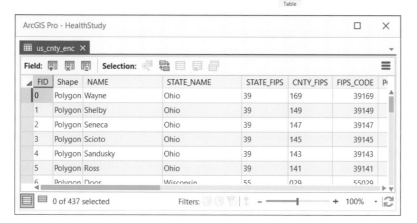

What is the field name that indicates the state within which the county features are located?

How many residents of Wayne County are between the ages of 22 and 29 years?

6. **Close the attribute table.** You'll want to limit the us_cnty_enc.shp dataset to the state of Illinois only. You might consider a few different operations to achieve this subset:

Definition query. You can set a definition query to limit the visible counties to only those within the state of Illinois. Definition queries are helpful when you want to work with a subset of data in a map while maintaining the source data. You can set a definition query on the Data tab of the Feature Layer ribbon. However, you must think ahead—what is next in the workflow? You want to append Illinois obesity statistics (which are in a stand-alone table) to a spatial layer of Illinois counties. You will do this by performing a join operation, which is explained in detail in exercise 3b. But with a definition query defined, the source data does not change, so the operation will attempt to join Illinois obesity rates to every county in the five-state dataset. Although this operation works, it takes a little longer, so it is best to go another route.

TIP *You will work with definition queries in chapter 5.*

Clip. You can use the Clip operation to select a portion of us_cnty_enc.shp based on another layer—namely, a layer that provides a boundary of the state of Illinois. However, you do not have an Illinois boundary layer now. There is another way.

TIP *You will work with the Clip tool in chapter 7.*

Select and export. You can select only those counties that belong to the state of Illinois, and then export the selected features to a new dataset. Then you will have an exact one-to-one correlation between your spatial attribute table (because the new output will have one record for each Illinois county) and your stand-alone table (obesity statistics). This method is the best for the current situation.

Select features by attributes

You will create a selection in which the state name attribute value equals Illinois.

1. **On the Map tab, open the Select By Attributes tool.**

    The Select Layer By Attribute tool opens in the Geoprocessing pane.

2. **For Layer Name or Table View, select the us_
    cnty_enc shapefile. For Selection type, maintain
    the default New selection (click the down arrow
    to see additional options).**

3. **Under Expression, click Add Clause, and then choose the correct options to create the following expression: STATE_NAME is Equal to Illinois.**

The expression you created is called an *attribute query*—a request for features in a table that meet user-defined criteria.

4. **Click Add to add the expression to the tool, and then click Run.**

A message window temporarily appears when processing is complete, and Illinois counties are highlighted in the selection color (aqua, by default).

Now you will create a new feature layer from the selected set of counties.

> **TIP** *You have made a selection based on attributes, but you can also make a selection based on location. For instance, suppose you want to select parcels that are within the boundaries of a flood zone. In this case, use the Select By Location tool. You can choose from 15 relationship types, including "Completely within," "Have their center in," "Intersect," and "Within a distance." You will use Select By Location in chapter 7.*

Export the selection to a new dataset

1. **In the Contents pane, select us_cnty_enc.**

2. **On the Data tab, click the Export button, and then choose the Export Features option.**
   Export
   Features

3. **In the Geoprocessing pane, for Input Features, select us_cnty_enc. For Output Feature Class, maintain the default location but change the feature class name to** Illinois_cnty.

   > **TIP** *To see the location of the output feature class, hold your pointer over the text box.*

Notice that the new output will be a geodatabase feature class, rather than a shapefile. By default, all processing outputs are saved to a project geodatabase—the conversion from shapefile to geodatabase feature class is done behind the scenes.

4. **Run the tool.**

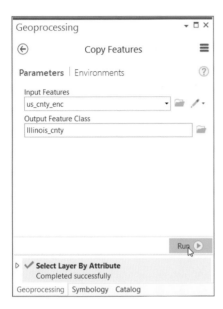

5. **Remove us_cnty_enc.shp from the map contents. Zoom to the extent of the Illinois_cnty layer. Change the map name to** Illinois.

REMIND ME HOW

- *Right-click layer and click Remove.*
- *Right-click layer and click Zoom To Layer.*
- *In the Contents pane, click Map once to select it and click again to make it editable. Type **Illinois**, and then press Enter.*

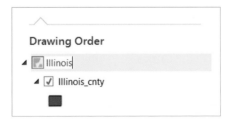

6. **If you want to change the color of the new layer, do so now:**

- **In the Contents pane, click the polygon symbol below Illinois_cnty to open the Symbology pane.**
- **In the Symbology pane, go to the current symbol properties so that you can modify the symbol manually.**
- **Click the Color selection down arrow to choose any color you like.**

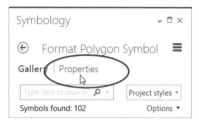

***TIP** If your Symbology pane is floating and you do not see the Color box, try dragging the divider bar to reveal the Appearance options.*

- **Click Apply, and then click the back arrow to return to the main Symbology pane. Close the pane if you want.**

Your preprocessing work is complete—you have created a project and added two datasets: a nonspatial table with obesity statistics for the state of Illinois and a spatial dataset that includes county demographic statistics for five states. Because you are interested in only the state of Illinois, you selected counties with a STATE_NAME field value of Illinois, and you exported these features to a new feature class (Illinois_cnty) stored in the project's geodatabase. In the next exercise, you will join the obesity statistics to the Illinois_cnty attribute table, so you can see the obesity data on a map.

7. **Save the project.** You will continue working with the same project in exercise 3b.

EXERCISE 3B

Incorporate tabular data

Estimated time to complete: 40 minutes

As you began researching obesity rates in Illinois, you quickly found a CDC website that offers reliable statistics on obesity prevalence. However, the data is offered in a spreadsheet table, and it cannot be displayed on a map because there is no coordinate or shape information. The data is broken down by counties, which is helpful. Using an *attribute join* operation, you can append one attribute table (the *join table*) to another attribute table (the *target table*), provided you have a common attribute field in each table—for instance, a feature name or numerical identifier. Then you can visualize, symbolize, label, and analyze the layer based on the new attributes. This process is what you will do here: using a county identification code as the common attribute, or *key*, you will join the CDC data to the Illinois county feature class.

> **TIP** *The common attribute field name itself does not have to be the same. It must be only the same type—for example, text fields are joined to text fields and numbers to numbers.*

Exercise workflow

- *Join CDC nonspatial data to the spatial layer of Illinois counties so that you can display the CDC data on the map.*
- *Symbolize the Illinois county layer to reflect the obesity prevalence statistics for a single year (2004).*
- *Copy the county layer, rename it, and import the symbology scheme used for the 2004 layer, but instead display the 2005 obesity prevalence statistics. Do the same for every year up to and including 2010.*
- *Use the Swipe tool for a quick visual comparison.*
- *Add a median income layer and look for data correlations.*

Join data tables

1. **In ArcGIS Pro, continue working with HealthStudy.aprx in the \\EsriPress\
    GTKAGPro\HealthStudy\Data folder.**

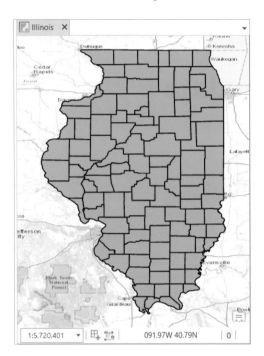

> **TIP** *If you did not complete or save your work from exercise 3a, open
> HealthStudy.aprx and add Illinois_cnty from \HealthStudy\
> Results\HealthStudy2.gdb and Obesity_Prevalence.dbf from the
> \HealthStudy\Data folder.*

2. **In the Contents pane, right-click Obesity_Prevalence and click Open.**

Tables are composed of rows and columns. Columns are often called *fields*. The column headers tell you what kind of information is provided for each feature represented by a row—in this case, the features are Illinois counties. Fields include OID (sometimes called Object ID), a unique identifier assigned to every row in an ArcGIS table; State, which shows the state name in which each county is located (in this table, all values are Illinois); FIPS_Code, a Federal Information Processing Standard (FIPS) code assigned to every US county; County, showing county names; and obesity statistics fields—for example, Number04 shows the number of people who are considered obese in the year 2004, Percent04 shows the percentage of the total county population that is considered obese in 2004, and so on.

3. **Scroll to the right of the table to view all of the fields.**

How many years of data are represented in the table?

PREPARING TABULAR DATA

Sometimes you must do some work to get tabular data in shape. When we first down-loaded this table from the Centers for Disease Control and Prevention website (www.cdc.gov), it looked like the graphic.

data_Illinois [Compatibility Mode]								
Obesity Prevalence							**2004**	
State	**FIPS Code**	**County**	**Number**	**Percent**	**Lower Confidence Limit**	**Upper Confidence Limit**	**Age-adjusted Percent**	**Age-adjusted Lower Confidence Limit**
Illinois	17001	Adams County	11970	24.4	20.3	29.2	24.3	20.0
Illinois	17003	Alexander County	1673	25.1	20.6	30.3	25.0	20.3
Illinois	17005	Bond County	3302	24.2	19.8	29.5	24.2	19.8
Illinois	17007	Boone County	8807	25.9	21.2	31.4	25.7	21.1
Illinois	17009	Brown County	1334	24.0	19.1	30.0	24.3	19.4
Illinois	17011	Bureau County	6008	23.1	19.1	27.7	22.9	18.7
Illinois	17013	Calhoun County	941.3	23.8	19.1	29.0	23.6	18.9
Illinois	17015	Carroll County	2982	24.3	19.8	29.5	24.1	19.4
Illinois	17017	Cass County	2418	23.9	19.3	29.3	23.8	19.2
Illinois	17019	Champaign County	30420	22.8	19.1	26.9	23.0	19.5
Illinois	17021	Christian County	6340	24.0	19.7	29.1	23.9	19.5
Illinois	17023	Clark County	3018	23.9	19.3	29.1	23.7	19.0
Illinois	17025	Clay County	2627	24.8	20.2	30.0	24.7	20.0
Illinois	17027	Clinton County	6462	24.0	19.6	29.2	24.0	19.6
Illinois	17029	Coles County	9013	23.1	18.9	28.0	23.4	19.2
Illinois	17031	Cook County	818500	21.5	20.2	22.8	21.4	20.1
Illinois	17033	Crawford County	3750	24.6	20.2	29.7	24.5	20.0

This table was originally formatted for easy reading; however, the table does not work if you try to use it in a table join operation in ArcGIS. The first row of a table must con-tain the field names that describe the values in their respective columns. So, we had to delete the first three rows and modify the headings in the fourth row. To maintain the descriptive table name "Obesity Prevalence" found in row 3, we incorporated it into the table name, replacing the original, ambiguous file name of data_Illinois. To maintain the years for which the data was collected, we added the appropriate year to each field name that contained statistical values. Then we truncated field names and replaced spaces with underscores (_) to maintain the database rules of no spaces and 10 characters. We also deleted columns that were not necessary for our study. Finally, we converted the table to an ArcGIS Pro–compatible .dbf file.

Remember, to join a spatial table and a nonspatial table, they must have at least one field in common—the field name need not match, but the values within the fields must correlate. In this case, the common field is the one that stores FIPS codes. Although the field names can be different, in this case they have the same logical name: FIPS_CODE.

4. **In the table, notice the field named FIPS_CODE. If you want, open the Illinois_cnty table and notice that it also has an FIPS_CODE field. When you are finished, close both tables.**

5. **In the Contents pane, select Illinois_cnty. On the Data tab, click the Joins button, and then click Add Join.**

The Add Join tool opens in the Geoprocessing pane.

In the Geoprocessing pane, the layer name should already be populated with Illinois_cnty.

6. **For Input Join Field, select FIPS_CODE. For Join Table, select Obesity_Prevalence. For Output Join Field, select FIPS_CODE.** Notice the warning symbol (the exclamation point in the yellow triangle) next to Input Join Field. It tells you that performance will be improved if you first index the table. You can ignore the warning in this case, but if you want to take the extra step, do the "On your own" exercise.

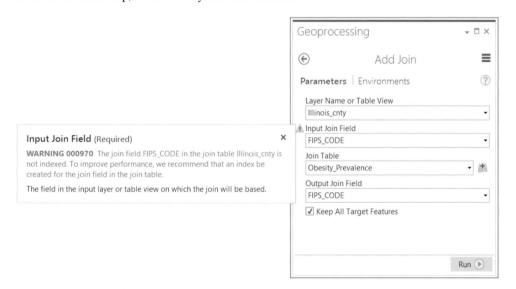

ON YOUR OWN

- *Open the Add Attribute Index geoprocessing tool.*

REMIND ME HOW

- *In the Geoprocessing pane, click the back arrow. Type index in the Search box, and press Enter.*
- *Click the first return—the Add Attribute Index tool from the Data Management toolset.*

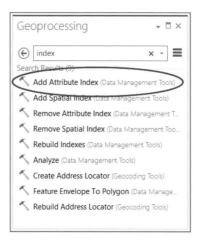

- *Choose the correct tool parameters (as shown in the graphic). The index name is required for geodatabase feature classes.*

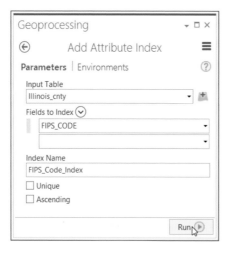

- *Run the tool. When indexing is complete, again open the Add Join tool and fill in the correct parameters as described in step 6.*

7. **Run the Add Join tool, and then open the attribute table for Illinois_cnty to see the operation's results. Scroll through all of the fields in the table.** Notice that when you pause over a field name, the table name is shown in parentheses, so it is easy to tell which attributes are from the original layer and which are from the joined table.

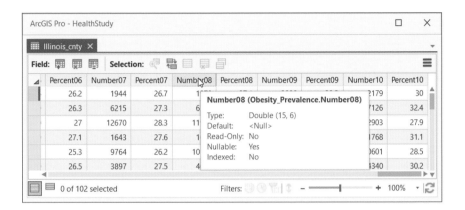

8. **Close the attribute table.** Joining data is a powerful GIS function. It is important to realize that the joined attributes remain with the layer you joined them to in the map; however, the underlying feature class is not modified. If you add another copy of the Illinois_cnty feature class to the map, you will not see the joined attributes in the new layer. In this way, join operations are considered temporary. However, if you save your work, the joined data will be there each time you open the project. If you want to create a dataset that permanently includes the joined attributes, you can now export the join layer to a new feature class with a new name, as you learned to do in exercise 3a.

Apply informative symbols

Now you are ready to display the joined information on the map. You will do so using *graduated colors*. The term *graduated colors* means that instead of using the same symbol for each feature, features are assigned a color that represents a quantity.

1. **On the Appearance tab, click the Symbology button's down arrow.**

   TIP *Make sure Illinois_cnty is still selected in the Contents pane to reveal the Feature Layer ribbon.*

Notice the symbology options (for more information, see the sidebar, "Feature symbol options" in chapter 2). You can choose categories if you want to display descriptions, such as a land-use category, or a value indicator such as Low, Medium, or High. In this case, you want to display counties using a quantity—namely, percentage value.

2. **In the Symbology drop-down box, click Graduated Colors.** ArcGIS Pro automatically applies a graduated color scheme based on the first value field it finds. You will change the value field to the one that contains the data that you are interested in.

3. **In the Symbology pane, click the Field down arrow, and select Percent04.**

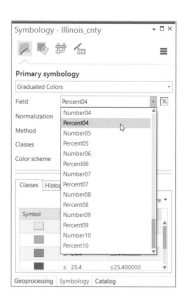

ArcGIS Pro by default chooses five categories based on a *natural breaks (Jenks) classification* method.

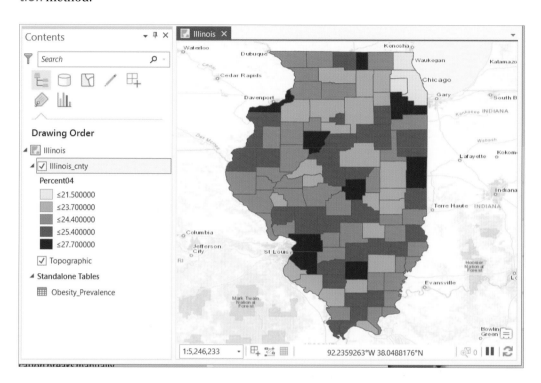

CLASSIFICATION METHODS

When mapping quantitative data using graduated colors or symbols, you want to classify the data appropriately so that the information is not skewed or misleading. ArcGIS has seven classification methods.

Manual interval classification

Using manual intervals, you can modify the classification breaks manually.

Equal interval classification

With equal intervals, the entire range of the data is equally divided by the number of classes you choose. If the range of values is 1 to 1,000 and there are five classes, this method creates classes of 1–200, 201–400, 401–600, 601–800, and 801–1,000. This method is useful for highlighting changes in the extremes and is most appropriate for familiar data ranges, such as percentages or temperatures. Using this method, it is also possible to have classes that contain no values.

Defined interval classification

A defined interval is similar to an equal interval. However, you can define the interval size to determine how many classes there are. If the students in your class have a range of test scores from 50 percent to 90 percent, for example, and the interval size is 10, this method creates four classes: 50–60, 61–70, 71–80, and 81–90.

Quantile classification

With quantiles, all classes have the same number of features. This method is appropriate when data is linearly distributed. It can create a balanced-looking map because no classes have too many, too few, or no values, but the resultant map can be misleading. The key is to increase the number of classes to prevent data from being placed in a different class. When an appropriate number of classes is used, the data can be visually represented more accurately.

 Natural Breaks (Jenks)
Numerical values of ranked data are examined to account for non-uniform distributions, giving an unequal class width with varying frequency of observations per class.

 Quantile
Distributes the observations equally across the class interval, giving unequal class widths but the same frequency of observations per class.

 Equal Interval
The data range of each class is held constant, giving an equal class width with varying frequency of observations per class

 Defined Interval
Specify an interval size to define equal class widths with varying frequency of observations per class

 Manual Interval
Create class breaks manually or modify one of the preset classification methods appropriate for your data

 Geometric Interval
Mathematically defined class widths based on a geometric series, giving an approximately equal class width and consistent frequency of observations per class.

Standard Deviation
For normally distributed data, class widths are defined using standard deviations from the mean of the data array, giving an equal class width and varying frequency of observations per class

Natural breaks (Jenks) classification

Natural-breaks classes are based on natural groupings inherent in the data, and boundaries are set in spots in which there are relatively large gaps between values. Developed by Professor George Jenks of the University of Kansas, this method is useful for classifying unevenly distributed data such as population.

Geometric interval classification

The geometric interval classification scheme creates class breaks based on class intervals that have a geometric series, such as a logarithmic distribution. This method was specifically designed to accommodate continuous data and minimize variance within classes to ensure that each class range has approximately the same number of values and that the change between intervals is fairly consistent.

Standard deviation classification

Data that you know is normally distributed (in a bell-shaped curve) can benefit from standard-deviation classification. This method creates classes according to a specified number of standard deviations from the mean value. A divergent color scheme is recommended for this type of classification.

You want to create your own categories that will be used to compare maps of every year for which you have data. If you are going to compare values, they must be broken into the same ranges or your comparison will not be helpful.

4. **In the Symbology pane, change the number of classes to 4. Choose the yellow-to-red color ramp.**

5. **Under Class breaks, click the last row's upper value twice to make it editable, replace the existing value with** 35, **and then press Enter.**

6. **Working from bottom to top, enter the following upper values for the remaining three symbol categories,** 31, 27, 23, **and then edit the symbol category labels as shown in the graphic.**

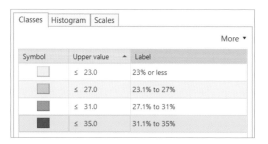

Notice that the new class breaks change your map dramatically: most counties are now in the second-lowest category, because no more than 27 percent of their populations are considered obese. No counties are in the highest category, which is reserved for counties with more than 31 percent of their population considered obese. The higher thresholds are required to accommodate the higher percentages of subsequent years, as you will soon see.

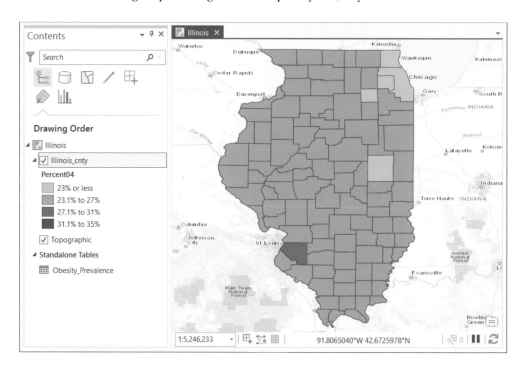

Import layer symbology

Next, you will create additional layers using the same symbology scheme that will reflect data from the years 2005 through 2010.

1. **In the Contents pane, right-click Illinois_cnty and click Copy. Then right-click Illinois (the map name) and click Paste.**

   Now you have two identical layers in your map.

2. **To keep the layers straight, rename the first one** 2004, **and the second one (the one at the top of the list)** 2005.

   If you simply change the symbol field to show 2005 percentage values, the breaks you defined earlier will not be maintained. Instead, you will import the symbology scheme from the 2004 layer but make it display the Percent05 values.

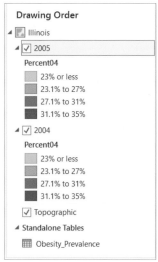

3. **With 2005 selected in Contents, go to the Symbology pane and click the** ☰ **Options button, and then click Import Symbology.**

The Apply Symbology From Layer tool opens in the Geoprocessing pane.

4. **Keep 2005 as the input layer. For Symbology Layer, select 2004. Ensure that the source field points to Obesity_ Prevalence.Percent04. Change the target field to Obesity_Prevalence.Percent05.**

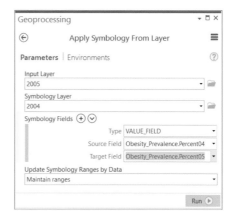

5. **Run the tool, and examine the results.**

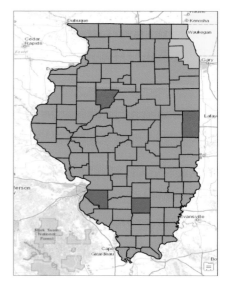

6. **Turn the 2005 layer on and off to see the difference between years.** In terms of map colors, in the 2005 map, two yellow counties have now advanced to light orange, and three light-orange counties have advanced to dark orange.

7. **Continue steps 1 through 6 five more times: paste the 2004 layer, rename it (the next one will be** 2006, **and then** 2007, **and so on), import the 2004 symbology scheme, and change the target field value to the appropriate PercentXX value for the input layer.**

    TIP *Make sure the input layer is the current layer you are working with and the symbology layer is set to 2004.*

8. **Turn off all of the obesity layers except 2004. One by one, turn on each subsequent year layer.**

    The visual impact of the data is striking. With a few exceptions, almost every county's obesity prevalence goes up with each year.

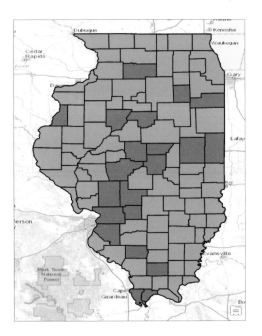

2006.   2007.

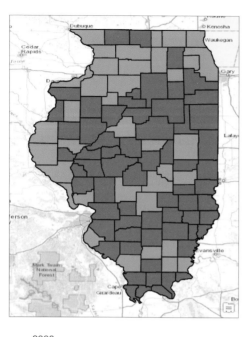

2008.

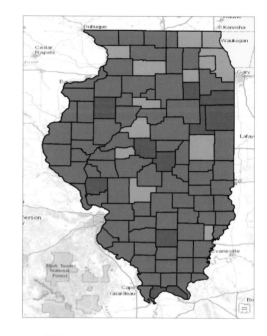

2009.

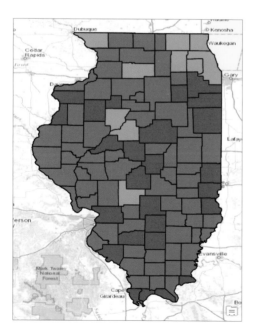

2010.

Use the Swipe function to compare layers

1. **Turn off and collapse all layers except 2004 and 2010.**

Drawing Order

▲ ☐ Illinois
  ▲ ☑ 2010
    Obesity_Prevalence.Percent10
    ☐ 23% or less
    ☐ 23.1% to 27%
    ☐ 27.1% to 31%
    ☐ 31.1% to 35%
  ▷ ☐ 2009
  ▷ ☐ 2008
  ▷ ☐ 2007
  ▷ ☐ 2006
  ▷ ☐ 2005
  ▲ ☑ 2004
    Obesity_Prevalence.Percent04
    ☐ 23% or less
    ☐ 23.1% to 27%
    ☐ 27.1% to 31%
    ☐ 31.1% to 35%

2. **Use the Swipe function to reveal the change over six years: with the 2010 layer selected in the Contents pane, click the Appearance tab, and then activate the Swipe tool.** 🖾 **Click at the top of the map, and press and hold the mouse button while dragging the arrow down to the bottom of the map.**

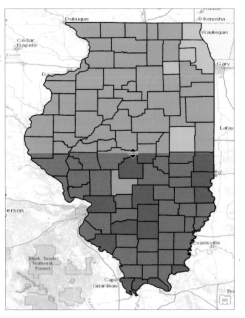

Overlay additional data

Some studies suggest that people with lower income have higher obesity and diabetes incidence. You want to see if there is an obvious correlation between median income and obesity rates in Illinois counties.

1. **Add IL_med_income.lyrx from the \HealthStudy\Data folder.** This is a *layer file*, which is a saved symbology scheme that points to a specific source dataset. In this case, the source dataset is a shapefile by the same name, stored in the same folder—IL_med_income.shp. You can create a layer file from any symbolized layer in ArcGIS Pro: simply right-click the layer in the Contents pane and click Save As Layer File. If you share a layer file, you must also share the source dataset along with it. You can create a *layer package*, which bundles the layer file along with the source data.

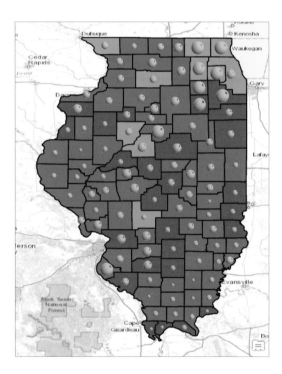

Do you see a clear correlation between income and 2010 obesity rates? Explain.

Although you cannot draw scientific conclusions from a visual observation, you can determine whether further analysis is warranted.

ON YOUR OWN

Create a layer file for the 2004 and 2010 layers. Save them to the project folder.

2. **Save the project.**

You are finished with this exercise. You will continue working with the same project for the next exercise.

EXERCISE 3C
Calculate data statistics

Estimated time to complete: 25 minutes

Your project is filling out. Now you can easily present maps that show obesity prevalence rates per county, with a different map for seven consecutive years. Your committee has requested a map that shows the percentage change over the seven-year study period. You do not have this information, but you can create it. In this exercise, you will add a new attribute field, and then populate the field with values. You will also calculate summary statistics for the state.

Exercise workflow

- *Add a new field to the Illinois counties layer that shows the percentage change between two years (2004 to 2010).*
- *Use the Field Calculator tool to populate the new field.*
- *Display the new attributes on the map, and modify the default labels.*
- *Use the Summary Statistics tool to create a table that shows the total number and average county percentage rate of obesity cases in Illinois as reported for the year 2010.*
- *Explore and configure the Infographics tool to get supplemental information about the study area.*

Add a new field

To display the percentage change between layers, you need a percentage change field in your attribute table. You will create that field now, and then you will populate the field values all at once using the Field Calculator function.

1. **Continue working in the HealthStudy project you have been working on in the previous two exercises.**

    TIP *If you did not complete the previous exercises successfully, you may use HealthStudy2.aprx, found in the \GTKAGPro\HealthStudy\Results folder.*

    Rather than add another layer to your already busy map, you will create a new map to display the percentage change values.

2. **In the Catalog pane, right-click Maps and click New Map.**

    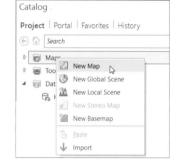

    A new topographic map is added to the project. Switch between the maps by clicking the tabs at the top of the map display.

3. **Name the map** Illinois2.

    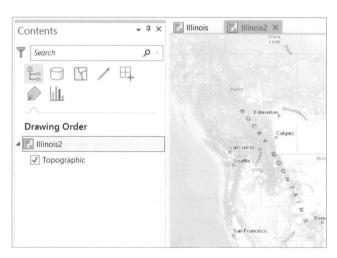

    TIP *To rename, either click the word Map twice, slowly, to make it editable, or double-click to open the properties and change the name on the General tab.*

4. **Click the Illinois tab to switch back to the original map.**

5. **In the Contents pane, right-click 2010 and click Copy.**

6. **Switch to the Illinois2 map, and in the Contents pane, right-click Illinois2 and click Paste.**

7. **Change the new layer name to** Percent change.

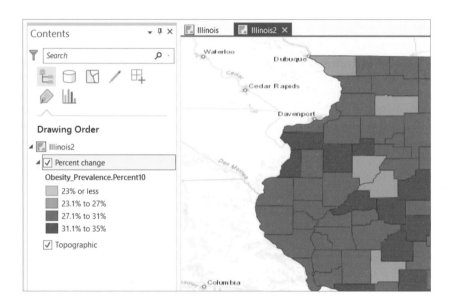

The layer name does not match what the layer is showing right now, but do not worry. You will fix that issue soon.

8. **In the Geoprocessing pane, find and open the Add Field tool, and complete these steps:**
   - **For Input Table, select Percent change.**
   - **For Field Name, enter** Perc_change.
   - **For Field Type, select Double (double precision).**
   - **Maintain the other default settings, and run the tool.**

TIP *As an alternative to searching in the Geoprocessing pane, you can open the Add Field tool from the geoprocessing tool gallery. On the Analysis tab, click through the gallery of tools until you find the Add Field tool under Manage Data. Click to open the tool. (If your application window is not maximized, you will see an Analysis Gallery button. Click it to access the gallery.)*

TIP *You can customize the gallery—tools can be added, removed, or reordered. (Removing a tool from the gallery does not delete it from the system— you can always find any tool using the Search function.) If your gallery is customized, it will look different.*

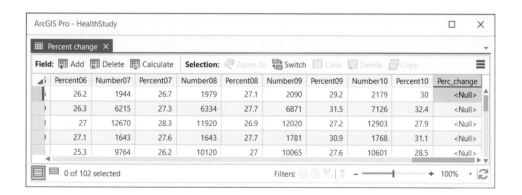

9. **When the process is complete, open the table for the Percent change layer, and find the new field.** It is the last field in the table. It currently has no values, which is represented by <NULL>.

You have added a permanent field to the source data of the Perc_change layer. Recall from exercise 3b that all of the obesity layers in this map come from the same base data (the Illinois_cnty feature class). They look different because their symbology schemes point to different attribute fields. If you open the attribute tables for the other obesity prevalence layers (2010, 2009, 2008, and so on), you will see all of the same attributes, including the new Perc_change field.

Calculate field values

1. **In the attribute table, right-click the new field and click Calculate Field.**

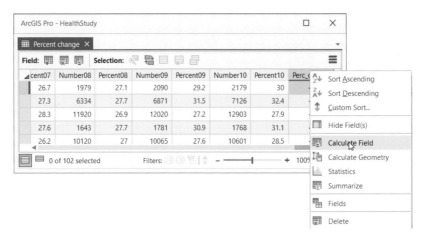

The Calculate Field tool opens in the Geoprocessing pane.

You want the new field to show the percentage change between 2004 and 2010. Because the percentage values already exist, your calculation is a matter of simple subtraction.

In the Calculate Field tool, the Input Table option should already point to Percent change.

2. **For Field Name, select Illinois_cnty_Perc_change.**

3. **Under Fields, double-click Percent10.**

4. **Next, click the subtraction symbol. Then double-click Percent04. Notice that the expression box fills in.**

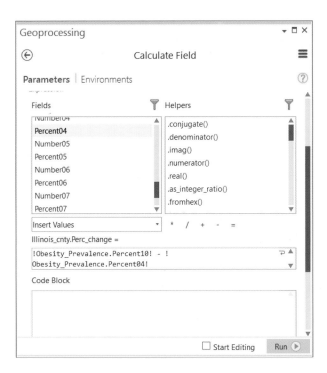

5. **Run the tool, and examine the new values in the table.**

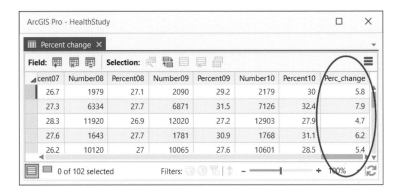

6. **Close the attribute table.**

Display a new field

TIP *Before the next step, you may need to save, close, and reopen the project.*

1. **Open the Symbology pane for the Percent change layer. Change the Field to Perc_change.**

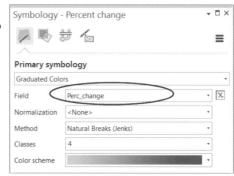

2. **Click the Advanced symbol options button.**

3. **Expand the Format labels category, and then change the number of decimal places to** 1.

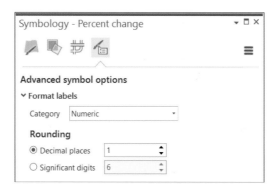

4. **Click the Primary symbology button to return to the main Symbology pane. If you like, change the color scheme to one of your choice.** A ramp, which progresses from light to dark colors, will communicate increasing percentage change most clearly.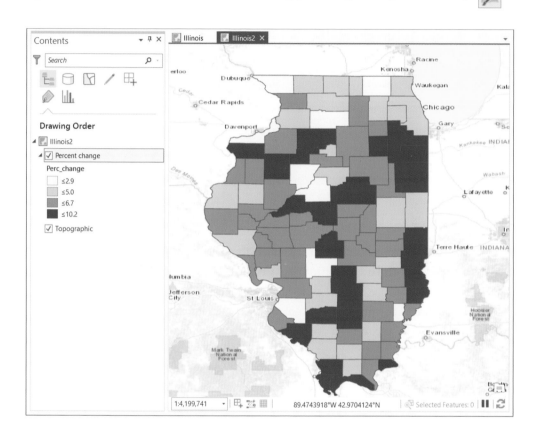

The Percent change layer shows, at a glance, which counties had the greatest increases between 2004 and 2010. These darker counties might be of interest to researchers and policy-makers. The reason for the sudden changes should be investigated.

Calculate summary statistics

Finally, you want to present a table that shows the most pertinent statistics at a glance.

1. **On the Analysis tab, click through the Analysis Gallery to find the Summary Statistics tool, and click to open it in the Geoprocessing pane. (Alternatively, search for Summary Statistics in the Geoprocessing pane.)**

2. **For Input Table, select the Percent change layer. For Output Table, maintain the default name, or choose a new one if you prefer.**

3. **For Field, click the down arrow and select Obesity_Prevalence.Number10. For Statistic Type, maintain the default setting (SUM).**

4. **In the second box that appears, select Obesity_Prevalence.Percent10. For Statistic Type, select MEAN.** The statistics that ArcGIS Pro generates will tell you the total number of obesity cases in Illinois in 2010, as well as the per-county average. You can also choose to calculate the same statistics for the other years represented in your project.

5. **Leave the optional Case field parameter blank, and then run the tool.**

   The output table is added to the bottom of the Contents pane.

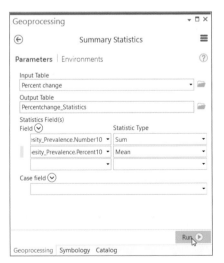

6. **Open and examine the table. Close it when you are finished.**

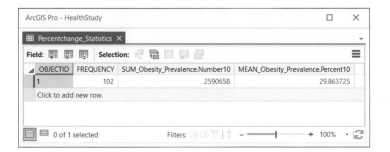

You can now present numbers for the entire state. As of 2010, Illinois had 2,590,658 obesity cases.

Examine infographics

1. **On the Map tab, activate the Infographics tool and click Alexander County at the southern tip of Illinois (see graphic).**

    TIP *The Infographics tool is enabled only if you are signed in to your ArcGIS Online organizational account.*

    The Infographics window shows you information about a feature that is not stored in the dataset; rather, it is "geoenrichment" data that is served from ArcGIS Online. The information is presented as statistics, charts, or graphs.

2. **Change the window options to show households by income, and compare the Alexander County setting so that the dots show a comparison to the entire state of Illinois.**

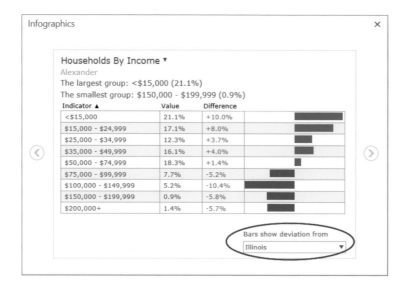

3. **Click the arrows on the sides of the window to navigate through the available data.**

What percentage of households have an income of less than $15,000 per year?

4. **Close the Infographics window when you are finished with it. Save the project.** You have done a great deal of research, discovery, and calculations for the Illinois counties health study project using ArcGIS Pro. Continue using this project for the next exercise.

EXERCISE 3D

Connect spatial datasets

Estimated time to complete: 15 minutes

In this short exercise, you will test one more hypothesis: that obesity rates correlate with higher numbers of "food deserts," or low-income areas with limited access to full-service grocery stores.

Exercise objectives

- *Create a relationship between a food desert layer (representing areas that are low income and have limited access to grocery stores) and the Illinois counties layer.*
- *Spatially join the layers so that you can generate a count of food deserts within each county.*

Relate tables

Next, you will *relate* two tables, which allows you to quickly see relationships between separate datasets.

1. **In ArcGIS Pro, continue working with the HealthStudy project, and activate the Illinois map. Turn off the IL_med_income and 2004 layers, and collapse their legends.**

   TIP *If you did not successfully complete the previous exercises, use HealthStudy2.aprx in the \HealthStudy\Results folder.*

2. **From the \HealthStudy\Data folder, add the IL_food_deserts.lyrx layer file. Open its table, and examine its attributes.**

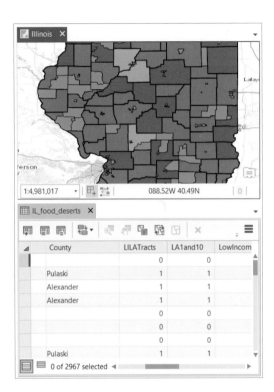

This layer shows Illinois data for the census tract level, rather than the county level. Notice the field named LILATracts. This field tells you if a census tract is considered a food desert (LILA stands for low-income, low-access). A value of 1 means that the tract is a food desert; a value of 0 means it is not. The symbology scheme is set to show only the tracts that have a LILATracts value of 1. Notice, too, that only rows that have a LILATracts value of 1 have a County value.

You will relate the IL_food_deserts table to the 2010 layer. That way, for each county, if there are statistics for food deserts, you can easily see them. You will relate the tables based on a common field: the County name field. Because only tracts that qualify as food deserts have a County name value, the tracts that are not considered food deserts are left out of the operation.

3. **Close the table.**

4. **In the Contents pane, right-click 2010, point to Joins and Relates, and click Add Relate. Select or enter the following tool parameters, and then run the tool:**

    - **Layer Name or Table View: 2010**
    - **Input Related Field: Illinois_cnty_NAME**
    - **Relate Table: IL_food_deserts**
    - **Output Relate Field: County**
    - **Relate Name:** Relate1
    - **Cardinality: One to many**

Geoprocessing   ▾ ▢ ✕

⊖   Add Relate   ≡

Parameters | Environments   ⑦

Layer Name or Table View
2010   ▾

Input Related Field
Illinois_cnty.NAME   ▾

Relate Table
IL_food_deserts   ▾ ⊞

Output Relate Field
County   ▾

Relate Name
Relate1

Cardinality
One to many   ▾

Run ▶

TIP *You are choosing a cardinality rule of "one to many," because each county may have more than one food desert.*

5. **On the Map tab, activate the Select tool, and click the county shown in the graphic.**

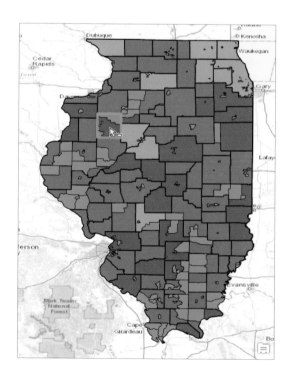

6. **Open the 2010 attribute table, and click the Show selected records button.**

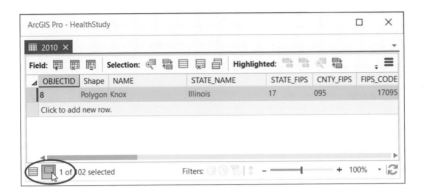

7. **At the top of the table, click the Options menu, then click Related Data >**
    Relate1:IL_food_deserts.

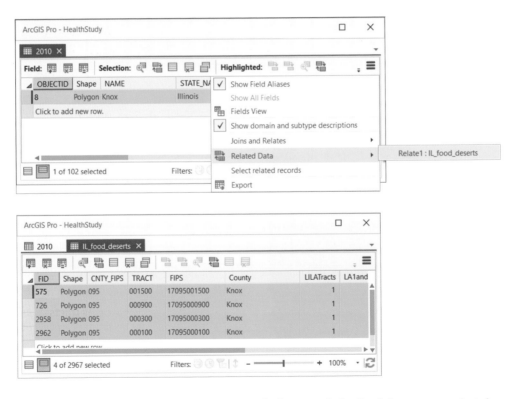

The IL_food_deserts table opens, showing only the records for food desert tracts that share a name with the selected record in the other table.

TIP *When multiple tables are open, you can reposition them, float them, or stack them.*

How many food deserts are in Knox County?

ON YOUR OWN

Select another county that appears to have multiple food deserts, and display the related records.

8. **Close the tables, and clear any selections.**

Spatially join data

Sometimes, rather than joining data based on a common attribute, you will want to join data based on location. This operation is called a *spatial join*. It allows you to define a spatial relationship between two layers (a target layer and a join layer) and combine their attributes in a new output layer. Depending on the types of data you want to join, you can choose how the spatial relationship is calculated. For instance, if you have county-level data but not state level, you can summarize county births for a state total, or you can average county income for a statewide average. You may also choose to generate a Count field that summarizes the number of features in the joined layer that intersect each feature in the target layer. Or you may generate a distance measurement between features in two layers. For example, you can define a relationship between city points and earthquake points based on distance, generating an output that shows the nearest city to each earthquake.

Here, you will spatially join the 2010 layer with the IL_food_deserts layer. The output will be a new geodatabase feature class, with the food desert count and affected population fields summarized for each county. Recall that LILATracts (low-income, low-access tracts) field values are either 0 or 1, with 1 indicating a food desert. By summarizing this field, you can quickly see how many food deserts are in each county. The POP_2010_1 field is the number of people who live in low-access areas. The field summary will show the total population of each county that may be affected by limited access to grocery stores.

1. **Turn off the IL_food_deserts layer.**

2. **On the Analysis tab, in the Tools gallery, click Spatial Join (from the Overlay Data toolset).**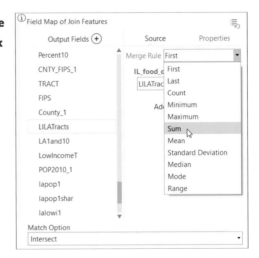

3. **Fill out the parameters as follows:**
   - **Target Features: 2010**
   - **Join Features: IL_food_deserts**
   - **Output Feature Class: default (\HealthStudy.gdb\c2010_SpatialJoin)**
   - **Join Operation: Join one to one**
   - **Keep All Target Features: checked**

4. **Under Field Map of Join Features, in the Output Fields list, scroll down and click LILATracts, and change Merge Rule to Sum.**

   Output Fields lists all attributes available from both target and join features. You can add or remove Fields, or change their merge rules or properties.

5. **In the same way, summarize the POP2010_1 field.**

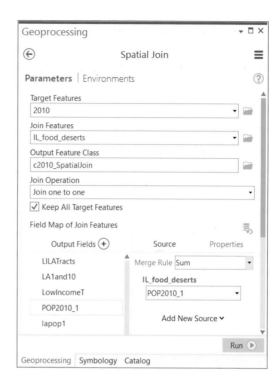

6. **Run the tool.** When the operation is complete, a new layer is added to the map.

7. **Symbolize the layer using a single symbol: No Background.**

   Next, you will label the layer based on the LILATracts field.

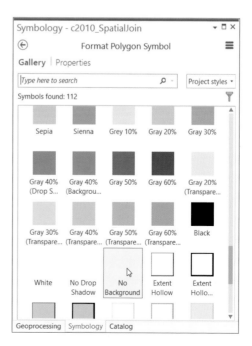

8. **With the Illinois_cnty_SpatialJoin layer selected in the Contents pane, click the Labeling tab. Set Field to LILATracts.**

9. **In the Contents pane, right-click the layer, and click Label to turn the labels on.**

TIP *Alternatively, click the Label button.*

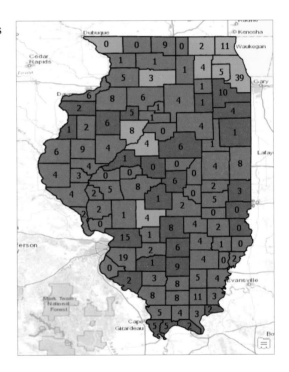

The labels show how many food deserts are in each county.

Although some counties have both high obesity prevalence and a high number of food desert tracts, there does not seem to be an obvious correlation at first glance. Some red counties have zero or only one food desert tract, while others have several.

Although you hoped to see some obvious correlative patterns between obesity, income, and food deserts in your initial visual analysis, the comparison of thematic maps is not the end of the story. Remember, you are mapping complicated problems that rarely have simple answers. Your next step would be to delve into more rigorous analysis techniques, using tools that model *regression analysis* to better compare your data variables. These advanced analysis methods are beyond the scope of this book, but if you are interested, for starters refer to the ArcGIS Pro Tool Reference topic: Tools > Spatial Statistics toolbox > Modeling spatial relationships.

Even if you find no clear statistical relationship between variables, you can either explore other possible variables or recommend that the committee not limit its target audience to people with a lower income, or even to people who live in a food desert. The committee might discuss advocating a large-scale education program delivered through various media to target individuals from all areas and walks of life.

10. Save and close the project.

Summary

This chapter is all about data relationships. You selected features by their spatial relationship to other features and then exported the selection to a new dataset. You turned nonspatial data into spatial data using a join operation that depended on an attribute relationship—a common field. You explored relationships within a single dataset when you calculated new fields and ran summary statistics. And you linked spatial datasets by relating their tables and performing a spatial join.

You should be getting comfortable using GIS to study problems and answer questions. In chapter 4, you will learn how to create and edit points, lines, and polygons.

Glossary terms

attribute query

attribute join

join table

target table

graduated colors

natural breaks (Jenks) classification

manual interval classification

equal interval classification

defined interval classification

quantile classification

geometric interval classification

standard deviation classification

layer file

layer package

relate

spatial join

## CHAPTER 4

Creating and editing spatial data

Exercise objectives

4a: Build a geodatabase

- Convert shapefiles to geodatabase feature classes
- Map x,y points
- Establish an attribute domain

4b: Create features

- Configure snapping options
- Create a line feature
- Enter attribute data

4c: Modify features

- Split polygons
- Merge polygons
- Modify lines and points
- Add map notes

You are probably getting comfortable exploring, querying, combining, and displaying spatial data. But what if you want to create it from scratch? This chapter teaches you how to create and edit spatial data.

Scenario: you work in the GIS department of Troutdale, Oregon. One of your jobs is to maintain the city's GIS database. You are considering converting all of the city's shapefiles into the geodatabase feature class format, because you want to take advantage of the benefits offered by the geodatabase. You will begin by converting just a few of the city's datasets so that you can decide whether it makes sense to convert the city's extensive data library. Then you will resume your maintenance duties as you create some new features and modify others.

WHY USE A GEODATABASE?

ArcGIS spatial data formats include raster data (such as surfaces, terrains, and imagery) and vector data in the form of shapefiles and geodatabase data, among other, less-common spatial data formats. Tabular data that records x,y locations and other attributes is also a type of spatial data that you can integrate into ArcGIS Pro, as you will see in the upcoming exercise. The most common data formats are the shapefile and geodatabase feature class. Both types of data group together points, lines, or polygons that represent geographic objects of the same kind, such as countries or rivers.

A *shapefile* is a simple, stand-alone data format. It stores geometry and attribute data for one set of features. Cities.shp, for example, might be a point feature class that represents cities. Shapefiles are composed of several files that have different extensions, although you will see only one item listed in the Catalog pane.

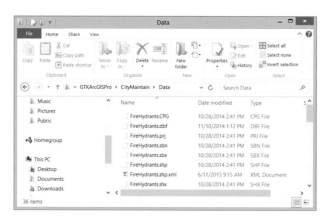
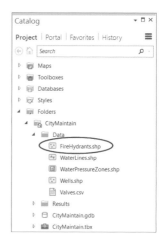

A shapefile's system files, and how it looks in ArcGIS Pro.

A *geodatabase*, by contrast, is a "storage container," in which sets of features are grouped into *feature classes*. The geodatabase can also store rasters and special geodatabase elements that facilitate capabilities that are not available with other data formats. For example, World.gdb might contain a polygon feature class of world countries, a line feature class of world rivers, a point feature class of world cities, and so on.

When working with ArcGIS Pro, you can add data from a variety of raster, vector, and tabular formats. When you add spatial data to ArcGIS Pro, you see it on the map, work with it as a layer, and usually are not interested in the format of the source file.

One of the advantages of a geodatabase is the ability to store multiple datasets, including rasters, which makes the geodatabase a great option for organizing data. You can group feature classes into a *feature dataset*—for example, sewer mains (lines) and manhole locations (points) can be grouped together in a Sewer feature dataset.

When shapefiles are imported into a geodatabase, the size on disk is dramatically reduced, so the geodatabase is an ideal model for data sharing.

Another advantage of using a geodatabase is the ability to create an attribute domain, which establishes and enforces valid values or ranges of values for an attribute field and minimizes data entry mistakes by prohibiting invalid values. For example, if "Open" and "Closed" are the domain values for a water valve status field, you cannot enter into the attribute table any values other than Open or Closed.

You can also create subtypes within geodatabase feature classes. A typical example is streets: for example, you can create subtypes for local, arterial, and collector streets. You can assign different default attributes or attribute domains to each; for example, local streets might have a default speed limit of 25 mph, whereas arterial streets might have a default speed limit of 35 mph. Or you can apply a different speed limit domain to each subtype.

Finally, the geodatabase can store behavior rules. Thus, a geodatabase can support specialized types of data, such as annotation feature classes, cartographic representations, geodatabase topologies, and networks, which all have specific behaviors. These special data types are increasingly important to GIS, although they are beyond the scope of this book.

DATA

In the \\EsriPress\GTKAGPro\CityMaintain\Data folder:

- FireHydrants.shp—a point shapefile that represents fire hydrants in Troutdale, Oregon *(Source: City of Troutdale)*
- WaterPressureZones.shp—a polygon shapefile that represents water pressure zones in Troutdale, Oregon *(Source: City of Troutdale)*
- WaterLines.shp—a line shapefile that represents water mains in Troutdale, Oregon *(Source: City of Troutdale)*
- Wells.shp—a point shapefile that represents wells in Troutdale, Oregon *(Source: City of Troutdale)*
- Valves.csv—a table that contains x,y, coordinate values for the locations of water valves in Troutdale, Oregon *(Source: City of Troutdale)*

EXERCISE 4A

Build a geodatabase

Estimated time to complete: 30 minutes

In this exercise, you will test the feasibility of switching the city's data collection to the geodatabase model. You will populate an empty geodatabase by converting shapefiles and mapping x,y location values found in a nonspatial table. All of the outputs will be placed in the geodatabase. Then you will create an attribute domain in the geodatabase, and apply it to the WaterLines feature class. This will ensure that valid values are entered into the table during the editing process.

Exercise workflow

- In a new project, convert shapefiles that represent various city water features into geodatabase feature classes.
- Convert a table that stores attributes and x,y location points (but no Shape field) into a point feature class.
- In the CityMaintain geodatabase, create a coded domain for the pipe diameter field so that entries are constrained to the standard pipe diameter measurements of four, six, eight, or twelve inches.
- Apply the domain to the WaterLines feature class.

Convert shapefiles to geodatabase feature classes

1. **In ArcGIS Pro, open CityMaintain.aprx, found in your \\EsriPress\GTKAGPro\ CityMaintain folder.**

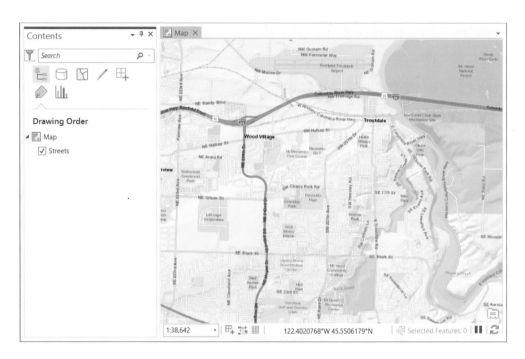

At this point, the map contains only a basemap layer (Streets).

2. **In the Catalog pane, expand Folders > CityMaintain > Data.** The project includes an empty geodatabase (CityMaintain.gdb), an empty toolbox (CityMaintain.tbx), and a data folder that has four shapefiles (.shp) and a table (.csv).

You want to convert these shapefiles into geodatabase feature classes in CityMaintain.gdb.

> **TIP** *Suppose you want to create a new geodatabase. To do so, right-click Databases in the Catalog pane, and click New File Geodatabase. Notice that you can also add an existing file geodatabase or create a connection to an enterprise geodatabase that resides on a server.*

3. **Open the Geoprocessing pane.**

REMIND ME HOW

ArcGIS Pro remembers the panes that you have opened before, so the Geoprocessing pane may already be open. If it is in the same space as the Catalog pane, just click the Geoprocessing tab. Otherwise, on the ArcGIS Pro ribbon, click the Analysis tab, and then click the Tools button.

You will search for a tool to help you convert the shapefiles into geodatabase feature classes.

4. **In the Search box of the Geoprocessing pane, type** convert. **Point to the Feature Class To Feature Class tool to read its ToolTip.**

This tool is the one you want.

5. **Click to open the Feature Class To Feature Class tool. For Input Features, add FireHydrants.shp. (You can either drag the shapefile from the Catalog pane or navigate to it in the \CityMaintain\Data folder.)**

6. **For Output, maintain the default setting (CityMaintain.gdb). Name the output feature class** FireHydrants**.** Notice that you can create an expression to convert a subset of the input feature class. You can also delete or modify certain fields when you convert. These settings are optional; you will not use them now.

7. **Run the tool. When the process is complete, keep the Geoprocessing pane open for now.**

8. **Expand the CityMaintain geodatabase in the Catalog pane to see the new feature class.**

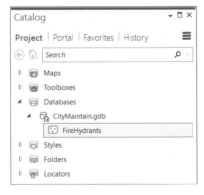

1
2
3
4
5
6
7
8
9
10

TIP *In the Catalog pane, you can see the CityMaintain geodatabase under Databases or under Folders > CityMaintain.*

The new feature class is also added to the map. Recall the symbol color is randomly assigned.

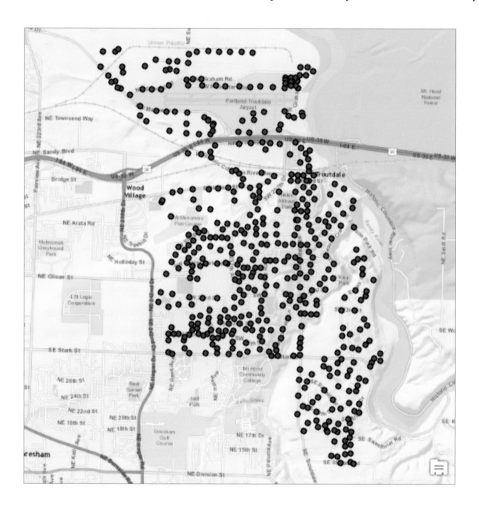

You still have three more shapefiles to convert. To avoid the tedium of running the tool three more times, instead you will use a script tool that converts multiple files at once. Whenever you find yourself repeating the same process many times, as in this data conversion, you should consider looking for a script, or consider writing a script yourself. Scripts automate repetitive processes. You will learn more about them in chapter 5.

9. **In the Geoprocessing pane, click the back arrow to return to the tool search. Point to the Feature Class To Geodatabase script tool to see the ToolTip.**

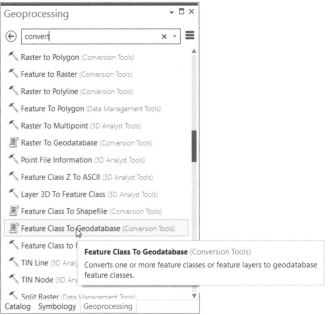

TIP *In the Geoprocessing pane, regular tools are denoted with a hammer icon;* and *script tools, which combine multiple processes, are denoted with a scroll icon.*

Notice that this tool allows you to convert several datasets at once. This tool is the one you want.

10. **Click to open the Feature Class To Geodatabase tool. For Input Features, add WaterLines.shp from the \CityMaintain\Data folder. (You can either navigate to it or drag it from the Catalog pane.) In the next input boxes that appear, add WaterPressureZones.shp and then Wells.shp.**

11. **For Output Geodatabase, browse to and select (or drag from Catalog pane) CityMaintain.gdb, and then run the tool.**

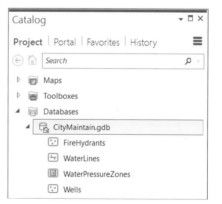

12. **In the Catalog pane, examine your new geodatabase feature classes. (You may have to refresh the database first—right-click it, and click Refresh.)** Notice that when using the script tool, the features classes are not automatically added to the map.

13. **Drag the new geodatabase feature classes (WaterLines, WaterPressureZones, and Wells) to the map. (Hint: use the Shift or Ctrl key to select them all at once.)**

> **TIP** *If you want to create a brand-new, empty feature class, use the Create Feature Class geodatabase tool. You can find it in the Data Management toolset, or launch it from the Catalog pane by right-clicking a geodatabase and clicking New > Feature Class. Then you can populate the feature class by using construction tools to create features.*

14. **Change the WaterLines color by right-clicking its symbol in the Contents pane and choosing a dark-blue shade from the color palette.**

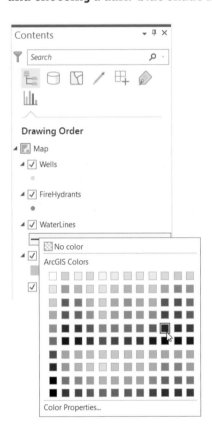

15. **Change the symbology for the other layers as indicated. To do so, click the current symbol under the layer in the Contents pane, which opens the Symbology pane. For each layer, choose the following symbol from the symbol gallery:**

- **Wells: Pentagon 1**

- **FireHydrants: Diamond 3**

- **WaterPressureZones: Black Outline (1pt)**

TIP *Point to the symbol to see the entire name, including line thickness.*

16. **Zoom to the Wells layer.**

REMIND ME HOW

Right-click Wells in the Contents pane, and click Zoom To Layer.

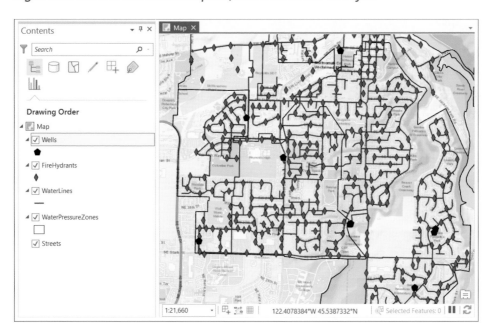

17. **Save the project.**

TIP *It is a good habit to save often to prevent having to redo your work in case of a software or hardware malfunction.*

Map x,y points

You also have a stand-alone table in your project that you want to convert to a feature class.

1. **Drag the Valves.csv table from the \CityMaintain\Data folder to the map and open it.**

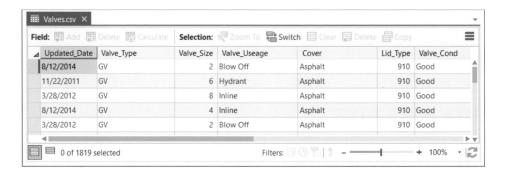

This table is nonspatial—that is, it does not have a well-defined geometry as feature classes do (there is no Shape field), so it cannot be displayed on a map. But the table contains spatial attributes in the form of x,y coordinates, which means that you can easily turn it into mappable data.

2. **Scroll all the way to the right side of the table to see the POINT_X and POINT_Y fields. Close the table when you are finished.**

3. **In the Geoprocessing pane, click Toolboxes. Expand Data Management Tools > Features, and read the ToolTip for the XY Table To Point.**

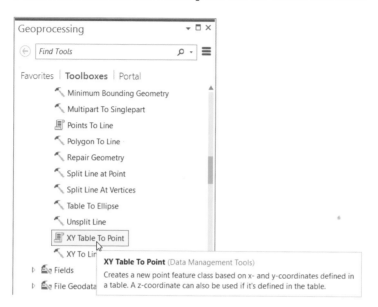

You will create a point feature class using the x,y coordinates stored in the table. You will make sure it is properly placed on the map by importing the projected coordinate system (see sidebar) of one of the other city layers.

COORDINATE SYSTEMS

All spatial data has a *coordinate system*, an important component of its *spatial reference*. The coordinate system defines features' positions on the earth's surface. Coordinate systems can be geographic or projected.

A *geographic coordinate system* uses latitude and longitude to define the locations of points on the surface of a sphere or *spheroid*. A *projected coordinate system* uses a mathematical equation (the *map projection*) to transform latitude and longitude coordinates into Cartesian or planar coordinates for display on a flat map. Particularly for detailed analysis or editing, it is important for all your layers to be in the same projected coordinate system.

ArcGIS employs *on-the-fly projection*, which means that it applies the projected coordinate system of the first layer added to all subsequent layers.

For more information, see the ArcGIS Pro Help topic: Maps and scenes > Map and scene properties > Coordinate systems > Coordinate systems, projections, and transformations.

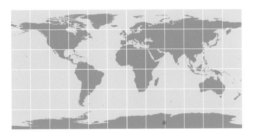

Left, a world map in the WGS84 geographic coordinate system, and *right*, in the Winkel-Tripel projected coordinate system.

---

4. **Open the XY Table To Point tool, and change these parameters:**
    - **For Input Table, click the down arrow, and select Valves.csv.**
    - **For X Field, select POINT_X.**
    - **For Y Field, select POINT_Y.**
    - **Leave Z Field blank.**
    - **For Coordinate System, click the down arrow, and select Wells.**

> **TIP** *To maximize learning, point to the information icons to read the tip for each tool parameter.* (i)

5. **Run the tool.** Don't worry about the warnings; this dataset does not have z-values (which record elevation or depth).

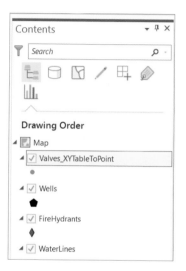

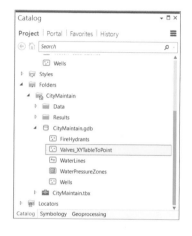

The feature class is added to the project geodatabase and to the map. (You may need to refresh the geodatabase in the Catalog pane to see it.)

6. **In the Contents pane, remove Valves.csv (the stand-alone table), and change the new layer name to** Valves.

7. **Change the Valves symbol to Circle 4, in a blue shade.**

REMIND ME HOW

In the Contents pane, click the default symbol for Valves. In the Symbology pane symbol gallery, select Circle 4 and click Properties. Change the color chip to a blue shade of your choice, and then click Apply.

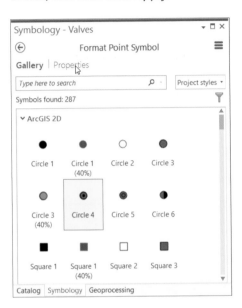 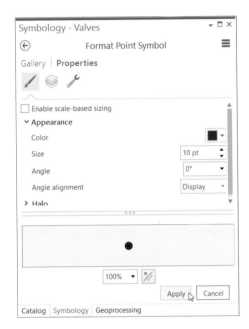

METADATA

Whenever you create new data, it is best to create *metadata*—textual information about the dataset—especially before sharing. To do so, right-click the dataset in the Catalog pane, and click View Metadata. A Catalog tab appears in the display area, and new options are available on the interface ribbon. On the Home tab, click the Edit tool. Enter your tags (which facilitate data searches), a summary of the dataset, and anything else you feel is necessary (such as data credits). When you are finished, click Save, and then close the Catalog pane.

Establish an attribute domain

To diminish the likelihood of data entry errors, you will set an *attribute domain* in your geo-database—a set of valid values, or a numerical range, to which attributes in each field must be limited. Specifically, you will create a coded value domain for the pipe diameter field of the WaterLines feature class. Pipe diameters must be four, six, eight, or twelve inches.

1. **Open the WaterLines attribute table. Click the Options button, and click Fields View.** ≡

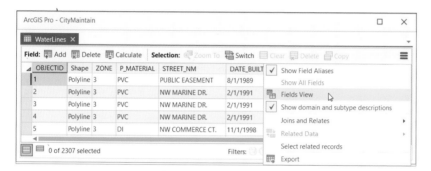

You now see the Fields tab, in which you can manage fields to make edits to a table's fields, modify their properties, delete fields, or create new ones. A Fields tab also appears on the main ribbon.

2. **At the bottom of the table, find the P_DIAM field. Select the cell in the Domain column. Click the cell again, and from the options select <Add New Coded Value Domain>.**

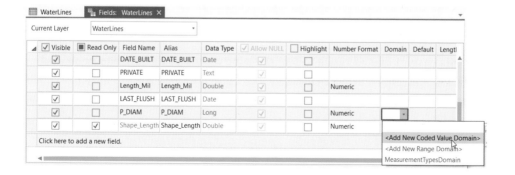

This step opens the Domains table. A Domains tab also appears on the main ribbon. You may notice a Measurement Types domain, created by default.

3. **Under Domain Name, enter** PipeDiam. **For Description, enter** Pipe diameter in inches.

4. **Maintain the default Field Type (Long), Domain Type (Coded Value Domain), and Split and Merge Policies (Default).**

5. **Under Code, type** 4, **and under Description, type** 4, **and then press Enter to create a new row.**

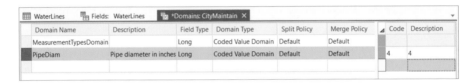

Domain Name	Description	Field Type	Domain Type	Split Policy	Merge Policy	Code	Description
MeasurementTypesDomain		Long	Coded Value Domain	Default	Default		
PipeDiam	Pipe diameter in inches	Long	Coded Value Domain	Default	Default	4	4

6. **Create three more Code/Description rows; and enter** 6, 6; **then** 8, 8; **and finally,** 12, 12.

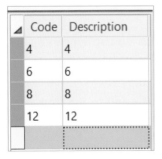

Code	Description
4	4
6	6
8	8
12	12

7. **On the Domains tab of the main ribbon, click the Save button.**

> **TIP** *In this case, it makes sense to use the pipe diameter as both the code and the description. In other situations, the code and description may be different. For instance, say that you want to establish a domain to constrain the soil type attributes within a study area. You can assign an alphanumeric code to represent each soil type, spelled out in the description column.*
>
> *For more about attribute domains, go to the ArcGIS Pro Help topic: Data > Geodatabases > Data design > Introduction to attribute domains.*

8. **When the domain is created, back in the Fields table, for the P_DIAM field, choose PipeDiam for the Domain column. Under Default, select 4 from the drop-down list, and then click Save on the Fields tab of the main ribbon.**

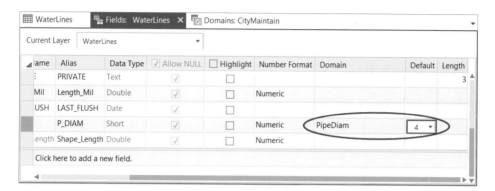

Now when you enter or edit waterline feature attributes, you will be constrained to enter pipe diameters of 4, 6, 8, or 12.

ON YOUR OWN

If you want, change the P_DIAM alias to Pipe Diameter.

9. **Save changes to the Fields table. Close the tables, and save the project.** You can continue to the next exercise, or close ArcGIS Pro and come back to it later.

EXERCISE 4B

Create features

Estimated time to complete: 30 minutes

The WaterLines layer is missing a line feature. You will create it in this exercise, using a feature template, construction tools, and editing options.

Exercise workflow

- *Configure the snapping options to facilitate editing.*
- *Create a new waterline that connects to existing water valve points.*
- *Enter attributes for the new waterline feature.*

Configure snapping options

Snapping is an editing option that acts like a magnet. When editing, if the point you create is within a specified distance of another feature's *vertex, endpoints, edge,* or *intersection,* it will jump (or snap) to coincide with another feature. Snapping allows you to accurately connect features, such as waterlines and valves, without impossibly precise sketching. You can enable various snapping configurations—for example, you might choose to snap to the endpoint of a line or the vertex of a polygon, or both.

To better decide which snapping options will benefit you, you will display the vertices of a feature that is next to one that you will be creating.

1. **Continue working in the CityMaintain project you worked on in exercise 4a.**

> **TIP** *If you did not successfully complete the previous exercise, open CityMaintain.aprx, add to the map all of the feature classes found in \CityMaintain\Results\CityMaintain2.gdb, and symbolize them as you want.*

First, you will change the selection and editing status of layers so you can more easily select and edit the features you want.

2. **On the Contents pane, click the List By Selection button. Turn off all layers except WaterLines.**

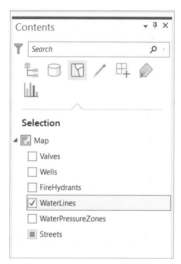

3. **On the Contents pane, click the List By Editing button. Again, turn off all layers except WaterLines.**

4. **Return to the Drawing order view of the Contents pane.**

5. **Go to the Water lines bookmark.**

Your dataset has two missing waterlines. Every valve should be connected to a waterline.

6. **Click the Edit tab on the ribbon.** Before you start creating, look at the existing line that runs along SW 19th Way.

7. **To see the composition of the line, click the Select tool on the Edit tab, and then click the line, which highlights in aqua.**

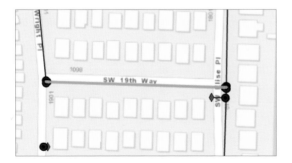

8. **In the Editor tool gallery, click the Vertices button to display the selected feature's edit sketch.** The Modify Features pane opens with the Edit Vertices tool active. The Modify Features pane shows the x,y location of each vertex of the selected feature.

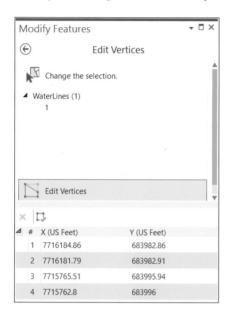

The Vertices toolbar appears on your map. It allows you to easily add or delete vertices, and finish or cancel an *edit sketch.*

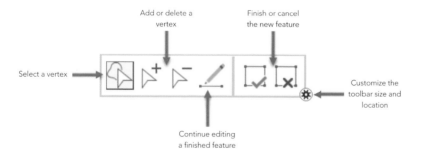

Refer to the Modify Features pane: How many vertices does the selected line have?

9. **Zoom to the corner of SW 19th Way and SW Wright Place to examine the selected line's vertices.**

> **TIP** *Here is some practical advice for detailed editing work: maximize your window, and zoom in closely. To pan or zoom the map without deactivating the current edit tool, use the mouse wheel shortcut for panning and zooming: scroll the wheel to zoom, press and drag the wheel to pan (or press C on your keyboard while dragging the map). Try it out—it is helpful when editing data.*

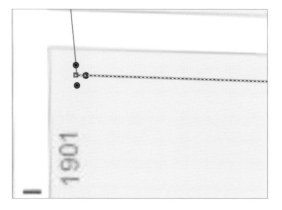

All lines have one vertex at each end and an unlimited number in between. By default, end-points are red squares; other vertices are green squares. In this case, there is a vertex placed at the point at which the line intersects a valve and one at either end of the line. (Turn Valves off and on to see better.)

10. **Clear the selection.**

11. **On the Edit tab, click the Snapping button to turn it on (when it is shaded blue, snapping is on).** You can configure the Snapping tool so that it snaps only to certain places, such as points, line ends, or polygon edges. When a snapping option is turned on, its icon is highlighted in the configuration options.

12. **Click the Snapping button down arrow, and configure the tool so that only Point Snapping and Vertex Snapping are turned on (again, blue shading indicates "on").**

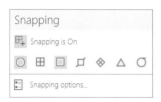

This configuration will allow you to snap to existing line vertices and valve points as you create your waterline features.

Create a line feature

You will create a new, L-shaped waterline that begins and ends at existing waterline junctions and intersects existing valves. The line you are about to create is selected (highlighted in aqua) in the graphic.

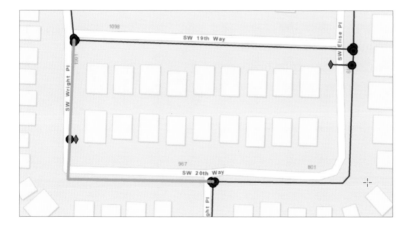

1. **On the Edit tab, click the Create button.** Create The Create Features pane opens. The Create Features pane shows the available *feature templates* you can use to create features. The templates match the layer names in the Contents pane by default. In this case, only one template is available, because you marked only one layer—WaterLines—as editable in the Contents pane.

   You can also choose which construction tools you want to show in the Create Features pane, or turn on a prompt for entering one or more attributes when the feature is drawn. To configure feature templates, right-click the template name, and click Properties. You should also realize that you are not limited to a single symbol for each feature template. If you have a layer that has graduated or unique symbols for certain feature types, all of the symbols will be available in the feature template. You choose the correct one based on the attributes of the feature you are creating.

   TIP *For more information, go to the ArcGIS Pro Help topic: Data > Edit geographic data > Feature templates.*

2. **In the Create Features pane, click the WaterLines feature template, and make sure the Line tool is active.** The Construction toolbar opens. This toolbar provides more tool options for creating line features.

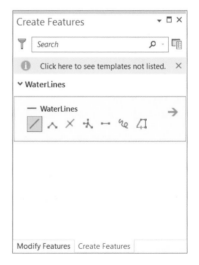

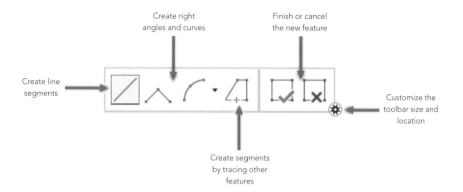

Create line segments

Create right angles and curves

Finish or cancel the new feature

Create segments by tracing other features

Customize the toolbar size and location

3. **Move your mouse pointer close to the junction of the two existing waterlines. Notice that when it gets close, a SnapTip (WaterLines: Vertex) appears.**

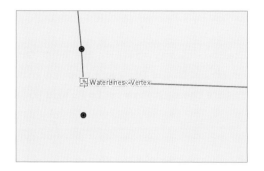

4. **With the SnapTip visible, click to place a vertex at the junction of existing waterlines.** Because of snapping, your click does not have to be so precise. If you see the SnapTip, the new vertex will be *coincident* with the existing vertex—that is, they are directly on top of each other.

5. **Move your mouse to the valve south of the point you just clicked. When you see the Valves: Point SnapTip, click to place another vertex.**

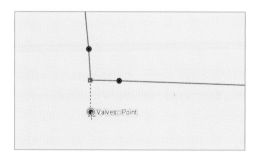

> **TIP** *If you place a vertex incorrectly, just click the Undo button at the top of the application.* ↺

6. **Zoom out, and pan south to the next valve. When you see the SnapTip, place another waterline vertex.** The next vertex you will place is where the waterline forms a right angle. There is no water valve to snap to, so you will use two edit commands to ensure proper placement: Distance and Deflection. You will constrain the next segment to a specific length of 58.5 feet, and you will ensure the straightness of the line by setting a deflection angle of 0 decimal degrees.

7. **Right-click anywhere on the map. On the menu, click Distance. In the Distance pop-up window, type** 58.5, **ensure the units are set to ft, and press Enter.**

8. **Right-click again. On the menu, click Deflection. In the Deflection pop-up window, enter an angle of** 0 **dd.**

Your next line segment is now constrained to the exact distance you entered, with no deflection from the previous line. The vertex is placed for you.

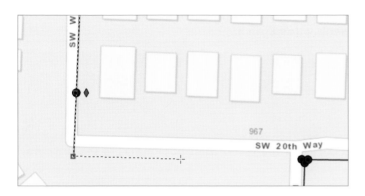

Just two vertices left.

9. **Extend the line parallel to SW 20th Way. Zoom in on the cluster of three valves. Snap to the first valve point, and click to insert a vertex.**

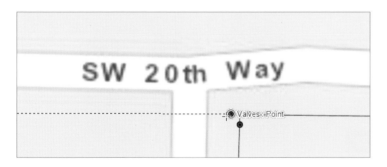

10. **Snap to the WaterLines vertex to place your final vertex.**

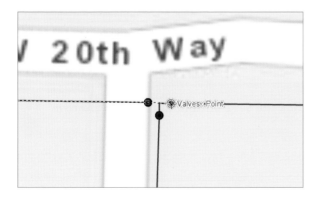

11. **Finally, move your mouse pointer slightly away from the last vertex placed, right-click, and on the context menu, click Finish.**

12. **Zoom out (or from the Map tab go to the Water lines bookmark) to see the completed new waterline.**

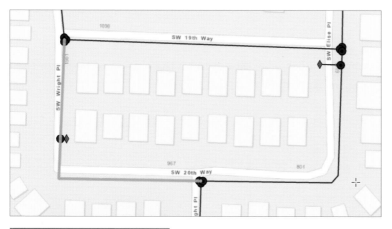

ON YOUR OWN

Create the waterline between the valve and fire hydrant on SW Wright Place. Try double-clicking to finish the sketch this time. When you are finished, clear the selection.

13. **Save your edits—on the Edit tab, click the Save button. Click Yes to confirm.**

Save

Enter attribute data

Now you will enter attributes for the new waterlines.

1. **Open the WaterLines attribute table, and show only the selected record.**

2. **Click twice in the cell below the ZONE field to make it editable. Type** 1, **and then press the Tab key to go to the next cell. In the P_MATERIAL field, type** DI **(for ductile iron).**

ON YOUR OWN

Enter these remaining attributes:

>*STREET_NM: SW 20th Way/Wright Place*
>
>*DATE_BUILT: 3/3/2015, 12:00 a.m.*
>
>*PRIVATE: N*
>
>*Length_Mil: 0.079449*
>
>*LAST_FLUSH: 4/15/2018, 12:00 a.m.*
>
>*P_DIAM: 8*

Recall that in exercise 4a, you created a coded domain rule for the P_DIAM field, and set the default value to 4. Notice that you are constrained to choose only a valid measurement: 4, 6, 8, or 12.

TIP *If you want to enter attributes before creating the feature, in the Create Features pane, click the blue arrow next to the appropriate feature template. Then enter attributes for the feature you are about to create.*

You have successfully created a new line feature, complete with attribute data.

3. **Save your edits, close the attribute table, and save the project.** You will continue working on the project in the next exercise.

GIS IN THE WORLD: MUNICIPAL GIS, FROM DESKTOP SOFTWARE TO WEB APPS

Riverside County, California, has been incorporating GIS into its operations since 1989. Its GIS team has developed GIS-enabled solutions, such as tracking fire damage, mapping bat locations (and rabies testing results), and maintaining city and county data, including transportation and cadastral data. It is now incorporating ArcGIS Online services into its business model to serve data and provide a space for sharing relevant maps and apps.

Read more about it here, "Riverside County Takes GIS to the Next Level": www.esri.com/esri-news/arcnews/spring14articles/riverside-county-takes-gis-to-the-next-level.

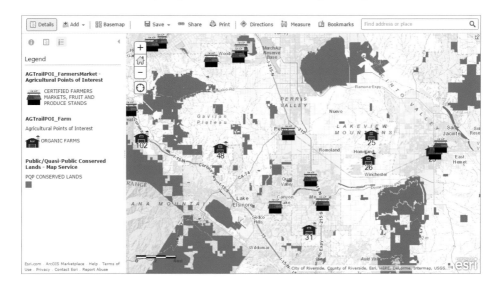

Using ArcGIS Online, you can find many map services and layers hosted by Riverside County. This map shows conserved lands, organic farms, and farmers' markets.

EXERCISE 4C

Modify features

Estimated time to complete: 35 minutes

Sometimes spatial data contains mistakes that must be corrected, and sometimes feature locations or boundaries change. In this exercise, you will modify existing features in the CityMaintain geodatabase.

To stabilize water pressure throughout Troutdale, two new water towers were built, resulting in changes to water pressure zones. You will split an existing zone into two zones and then merge one of the new, smaller zones into an adjacent zone.

Exercise workflow

- *Split a water pressure zone into two separate features.*
- *Merge two water pressure zones into a single feature.*
- *Move an erroneously placed waterline.*
- *Permanently delete a well.*
- *Use the Map Notes function to mark places at which edits have recently been made on the map.*

Split polygons

1. **Continue working in the CityMaintain project.**

    TIP *If you did not successfully complete exercises 4a and 4b, open CityMaintain.aprx, add all of the feature classes from \CityMaintain\ Results\CityMaintain2.gdb, and symbolize them as you want.*

2. **Zoom to the WaterPressureZones layer.**

3. **In the Contents pane, in the Drawing Order view, turn off Valves, FireHydrants, and WaterLines.** Only the Wells and WaterPressureZones layers are visible.

4. **In the Selection view of the Contents pane, make Wells, WaterLines, and WaterPressureZones the only selectable layers.**

5. **In the Editing view of the Contents pane, make the same three layers editable.**

6. **Return to the Drawing Order view of the Contents pane.**

7. **On the Map tab (or the Quick Access Toolbar, if you placed it there), click the Explore tool, and then click the water pressure zone shown in the graphic to identify it. Take note of the OBJECTID, ZONE, and Shape_Area attributes.**

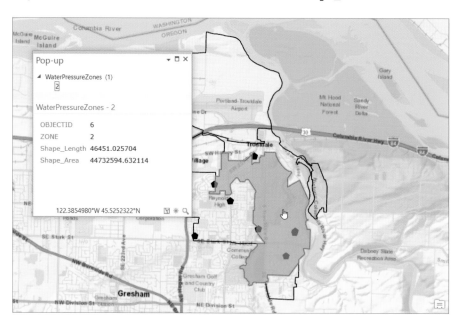

This is the feature you are going to split in two.

8. **Close the pop-up window.**

9. **On the Edit tab, check your snapping options to verify that the point and vertex snapping options are turned on.** If you have not made changes since the previous exercise, they should be on; ArcGIS Pro recalls your previous settings.

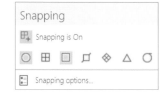

10. **Select water pressure zone 2, and then click the Split tool in the Tools gallery.**

TIP *Alternatively, if necessary, click the Modify button to open the Modify Features pane, and select the Split tool from there.*

11. **Zooming in as necessary, snap to the vertex shown here (indicated with a red circle in the graphic), and click to place a coincident vertex.**

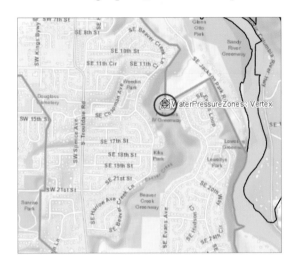

12. **Place a second vertex in the approximate location shown here.**

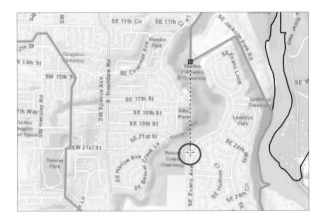

13. **Finally, snap to the vertex shown here, and double-click to place a final vertex and complete the split.**

TIP *You can finish a sketch several ways: double-click, press F2 on your keyboard, right-click and click Finish, or click the Finish button on the pop-up toolbar.*

One polygon feature has become two. You may notice when the split is complete that each new feature flashes briefly.

14. **Use the Explore tool to identify each water pressure zone. Notice the updated attributes.** The northernmost zone has a new OBJECTID value but has maintained the ZONE attribute of 2.

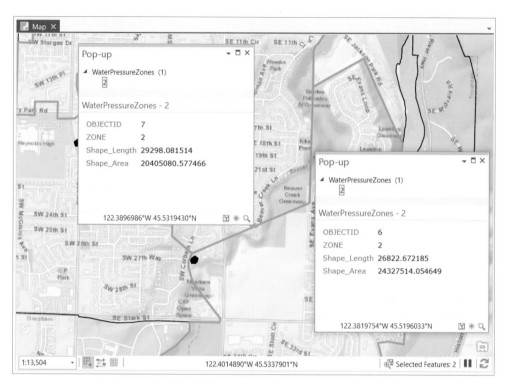

TIP *Your Shape_Area values may be slightly different. Your OBJECTID values may also be different, depending on your editing history.*

What happened to the Shape_Area value of the original water pressure zone?

15. **Clear the selection, and save your edits.**

ON YOUR OWN

There are now three polygons identified as zone 2. Edit the WaterPressureZones attribute table, giving the northernmost of the new polygons a new ZONE value of 6.

Merge polygons

Next, you will merge one of the new water pressure zones (the southernmost result of the split operation) with the smaller one below it. The zones you will merge are indicated by the red circle in the graphic.

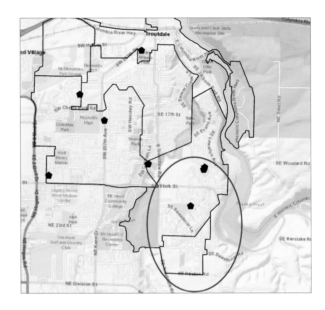

1. **On the Edit tab, in the Tools gallery, click the Merge tool.**

    TIP *Another way to find edit tools: open the Modify Features pane (by clicking the Modify button), and select All Tools.*

    The Merge tool prompts you to select two or more features.

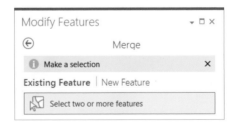

2. **Zoom into the polygons to be merged. Using the Select tool, click one of the polygons, and then press and hold the Shift key while you click the other.** Now both water pressure zones are selected.

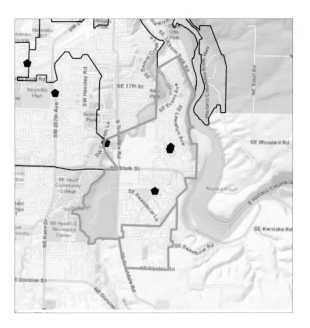

TIP *Notice that the Select tool can be found on both the Map and Edit tabs.*

The Modify Features pane is populated with information regarding the two selected zones. It tells you that you are about to merge zones 2 and 6.

3. **Under Features to merge, choose to preserve the attributes of zone 6. Click 6 to highlight the zone on the map.** This step is a good visual confirmation of what you are about to do.

4. **Click Merge to run the operation.**

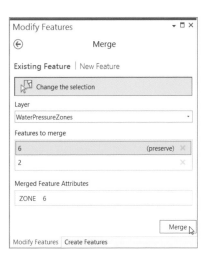

5. **Using the Explore tool, display the attribute pop-up window for the new larger water pressure zone.**
Notice that the new larger polygon has kept the ZONE value of 6. The Shape fields are automatically updated to reflect the new geometry of the now larger polygon.

6. **Clear the selection, and close the pop-up window.**

Modify lines and points

A waterline was placed incorrectly. You will fix the error by moving its vertices.

1. **Turn on the WaterLines layer.**

2. **Go to the Move bookmark, and reveal the vertices for the waterline shown in the graphic.**

REMIND ME HOW

On the Edit tab, activate the Select tool, click the line feature, and then click the Vertices tool.

The waterline runs alongside SE Pelton Avenue.

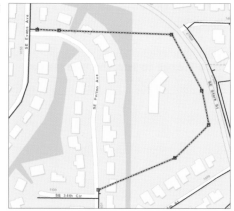

3. **One at a time, place your mouse pointer over each vertex, as circled in the graphic, and drag it to the approximate location indicated by the arrows.**

4. **Finish the modification by pressing F2 on your keyboard.**

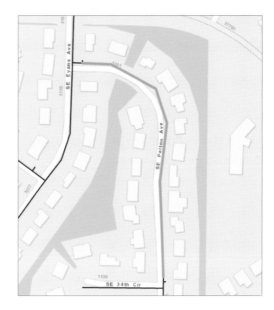

5. **Clear the selection.** Finally, you will remove a well feature that has run dry. Deleting features cannot be any easier—simply select a well, and click Delete.

6. **Zoom out slightly to see the well located in Sweetbriar Park.**

7. **Select the well feature on the map, and click the Delete tool or press Delete on your keyboard.**

✕
Delete

Saving your edits will permanently delete the well from the database.

ON YOUR OWN

Delete the waterline that led to the well.

8. **Save your edits and the project.**

Add map notes

You want a quick visual reminder of the changes you recently made to the database. You will use the Map Notes feature to create graphic features that highlight the new waterline and water pressure zone boundaries.

1. **Go to the Water lines bookmark.**

2. **On the Insert tab, in the Layer Templates group, click the Bright Map Notes button.**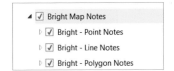

   A new group layer is added to the Contents pane.

3. **In the Contents pane, expand the Bright Map Notes group layer.** The group layer contains three notes layers for points, lines, and polygons. There are also three new feature classes in the CityMaintain geodatabase.

4. **If necessary, open the Create Features pane.**

5. **In the Create Features pane, click the Thick Line 1 symbol in the Bright-Line Notes layer, and make sure the Line tool is active.**

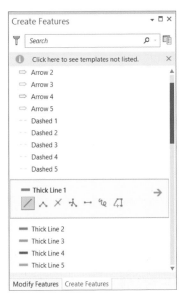

6. **On the map, draw two lines over the new waterline features, as shown here. Give the feature three vertices: one at the beginning, one at the angle, and one at the end.** Do not worry about snapping or exact placement; this is only a graphic, not a waterline feature.

7. **Finish the sketch, and then clear the selection.**

8. **Zoom to the extent of the WaterPressureZones layer, and turn off WaterLines for easier visibility.**

9. **In the Create Features pane, click the Polygon 1 symbol in the Bright-Polygon Notes layer, and make sure the Polygon tool is active.**

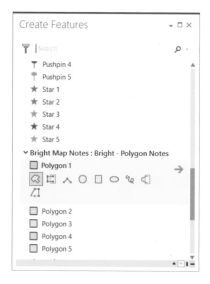

10. **On the map, draw a simple polygon that roughly corresponds to the new water pressure zone that resulted from splitting and merging old zones. (Again, it does not need to be exact.) When the sketch is complete, clear the selection.**

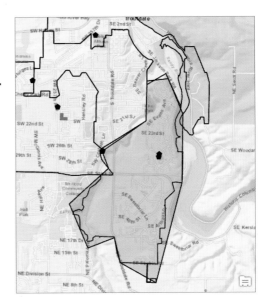

ON YOUR OWN

Make map notes to highlight the places at which you modified the waterline error and deleted the well.

11. **Go to the Troutdale bookmark. Turn on all layers.**

ON YOUR OWN

Set a visibility scale for the Valves layer so that the valves are only visible at scales larger than 1:10,000.

REMIND ME HOW

- *In the Contents pane, right-click Valves, and click Properties.*
- *On the General tab, under Out Beyond (minimum scale) select 1:10,000.*

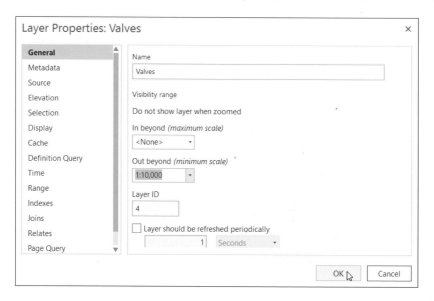

12. **Save your edits and the project.** You are finished with this chapter. If you are not continuing to chapter 5, you can close ArcGIS Pro.

Summary

Although a great deal of spatial data is circulating in the world, countless scenarios exist in which features must be deleted, modified, or created from scratch. You now have a foundational skill set for such tasks. In this chapter, we did not use many additional edit tools—for instance, we did not create any curved geometry—but now you should feel comfortable enough in the editing environment to experiment on your own. Remember, all edits are temporary until saved, and you can use the Undo button to fix mistakes, so do not be afraid to develop your skills through trial and error.

Glossary terms

shapefile

geodatabase

feature class

feature dataset

coordinate system

geographic coordinate system

spheroid

projected coordinate system

map projection

on-the-fly projection

metadata

attribute domain

snapping

vertex

endpoints

edge

intersection

edit sketch

feature template

CHAPTER 5
Facilitating workflows

Exercise objectives

5a: Manage a repeatable workflow using tasks

- Set up a project

- Proceed through preconfigured tasks

- Author a task

5b: Create a geoprocessing model

- Define the data

- Add operations to ModelBuilder

- Fill out the tool parameters

- Run the model

- Convert a model to a geoprocessing tool

5c: Run a Python command and script tool

- Define the data

- Execute a command using Python

- Use a custom script tool

- Package the project

You are now gaining a solid understanding of how you can use GIS to manage spatial data and solve spatial problems. In this chapter, you will learn about three ways in which you can store and automate multiple-operation workflows, an important capacity of ArcGIS software. Workflow definition, storage, and automation facilitate and standardize geospatial solutions. These capabilities are especially useful for organizations that anticipate performing a workflow many times. And you can change these workflows, including the input data and other parameters, as needed with each execution. Think of the workflow processes that you will learn about in this chapter as flexible, smart, "living" documentation.

First, you will learn about using tasks, a new feature introduced in ArcGIS Pro. This feature allows you to walk the user through a series of predefined steps that can incorporate (or call) commands and geoprocessing tools, including model and script tools. A task item might capture an entire workflow or one piece of a more complex solution. Tasks promote best practices, thereby improving efficiency and quality. They can serve as a workflow tutorial—as a documented series of instructions that call forth the correct commands or tools. And the tasks can be shared as part of a project, allowing users to share their knowledge.

Second, this chapter will introduce you to ModelBuilder, a geoprocessing environment that allows you to easily link one tool to another and run a set of operations one after another with the click of a button. ModelBuilder provides a helpful visual diagram of geoprocessing workflows and includes advanced capabilities such as looping and "if-then" scenarios (but we will not get that far in this book).

Finally, you will begin to learn how to automate your work with Python, the scripting language that is compatible with ArcGIS software. You will write a command and use a custom script tool (created with Python) that executes multiple operations.

Tasks, models, and Python scripts are often used together. For example, a task step might entail running a geoprocessing tool, which might be a custom script or model tool built using Python or ModelBuilder, respectively. It is helpful to understand all three environments, at least minimally, so that you can determine how to incorporate them into your own workflows. You can use the three environments to reproduce your workflows easily and share them with your organization and the ArcGIS community.

Scenario: you are employed by a nonprofit agency whose primary objective is to raise awareness of and develop solutions for international human rights violations. Taking advantage of the public information offered by the Armed Conflict Location and Event Data (ACLED) project, you are creating maps that convey the impact of political violence in developing countries. You will present your maps at an international human rights conference for eventual incorporation into an awareness campaign.

The ACLED dataset you are using provides information on political violence in developing countries from 1997 to 2013. Attributes include dates of conflict events, key actors involved (such as military forces, protesters, rebel groups, civilian protesters, and others), and event classification (such as riots and protests, battles, nonviolent activities by conflict actors, violence against civilians, and more). Your initial project focuses on the African continent. Because you intend to make many similar maps for various African countries, you are exploring automation capabilities in ArcGIS Pro.

GIS IN THE WORLD: MAPPING THE WALKABILITY OF CANADIAN NEIGHBORHOODS

A community designed to promote more walking and bicycling and less driving also benefits citizens' health and the environment. The Halton Region's Planning Services Division, along with the region's Health Department, spearheaded a GIS project to measure, quantify, model, and map the walkability of a large region of Ontario, Canada. They used ArcGIS and the ArcGIS® Network Analyst extension for data management and analysis. They also used ModelBuilder to automate the analysis so that it could be easily modified and repeated for other regions.

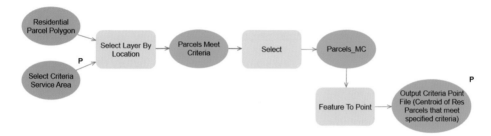

The project uses several models. This is an example of a model for selecting residential parcels within walking distance to a specified diverse-use (DU) development.

Read about the project here, "Modeling Walkability": www.esri.com/news/arcuser/0112/modeling-walkability.html.

DATA

In \\EsriPress\GTKAGPro\Conflict\Conflict.gdb:

- ACLED_2005_2018_Africa—a point feature class that displays locations of armed conflicts throughout Africa from 2005 to 2018 *(Source: ACLED)*

To get the data in its current state, the tabular data was downloaded from www.acleddata.com to an Excel file. Rows of data recorded prior to the year 2005 were deleted; the spreadsheet was saved as a .csv file and converted to a point feature class using the Table To Point tool in ArcGIS Pro.

EXERCISE 5A

Manage a repeatable workflow using tasks

Estimated time to complete: 30 minutes

In this exercise, you will begin using the Tasks pane, an ArcGIS Pro feature that allows you to follow or define a workflow for yourself or other users. A task item is a predefined series of steps, grouped into tasks, that walk you through a GIS workflow. Each *task* contains one or more related steps. Tasks are helpful to standardize business operations and promote best practices for a repeatable workflow. As you will see, some of the steps can be automated or semiautomated, saving time and reducing the possibility of user error.

Exercise workflow

- *Add the ACLED data for the African continent.*
- *Examine the dataset's attributes.*
- *Follow preconfigured tasks to do the following:*
  - *Limit the layer's definition so that it shows only conflicts in a single country during specific years.*
  - *Rename and symbolize the defined layer.*
  - *Select only those conflicts classified as "violence against civilians."*
- *Create a new task to do the following action:*
  - *Summarize the number of fatalities within the selected set of features.*

Set up a project

1. **In ArcGIS Pro, open Conflict.aprx from the \GTKAGPro\Conflict folder.**

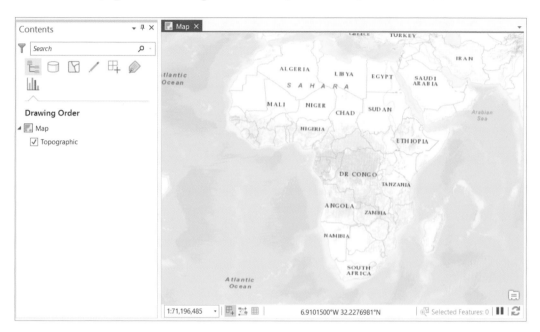

So far, the map contains only a topographic basemap layer, centered on the African continent.

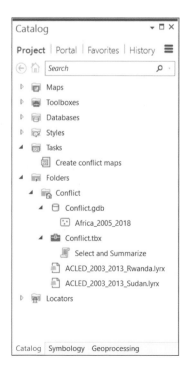

2. **Examine the Catalog pane by expanding the elements shown in the graphic.** The project includes two layer files, a geodatabase (Conflict.gdb) that contains a point feature class, a toolbox that includes a custom script tool, and a task item named "Create conflict maps."

First, you will add your project data to the map.

3. **Drag ACLED_2005_2018 from the Catalog pane to the map.**

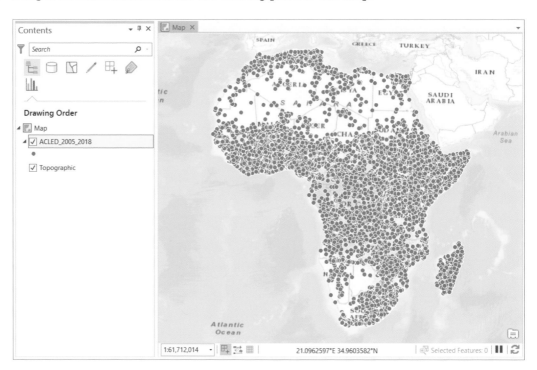

4. **Open the layer's attribute table, and examine the field values.**

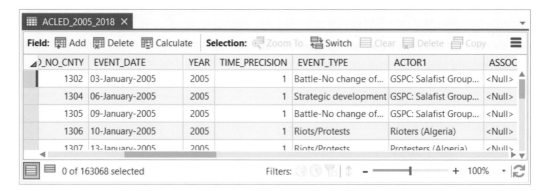

Can you name two types of conflict events that are recorded in this dataset?

---

5. **Close the attribute table.** Next, you will follow a task to perform a repeatable workflow.

Proceed through preconfigured tasks

1. **On the View tab, click the Tasks button to reveal the Tasks pane.**

When the Tasks pane opens, you will see a task item named "Create conflict maps." The task item includes three separate tasks: Create subset of features, Symbolize layer, and Make selection.

2. **Point to the first task to reveal the blue arrow. Click the arrow to open the task.**

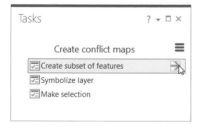

The first task contains three steps: the Progress indicator at the bottom says that you are on step 1 of 3. Familiarize yourself with the layout of the Tasks pane.

The first step facilitates the creation of a *definition query*, which limits the display of features to features that meet user-defined criteria.

3. **Read the step and instructions for step 1 of the task.**

   TIP *If necessary, expand the Tasks pane to see all of the instructions.*

4. **Click Run.** This step opens the query builder, which resides on the Definition Query tab of Layer Properties. The rest of the step relies on manual input, the instructions for which are documented in the task step.

5. **Carefully follow the step instructions in the Tasks pane to build the** *query expression*. **(Hint: you may need to click More in the Values box to see Sudan.)**

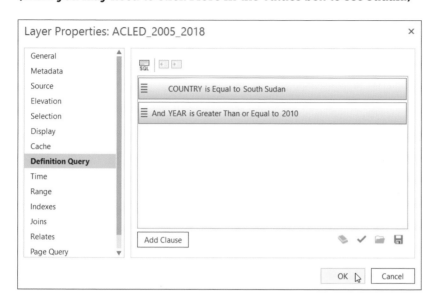

After the definition query is executed, the map displays fewer features.

6. **After you have closed the query builder, click Next Step to proceed to step 2: Rename Layer.**

7. **Click Run to open the layer properties, follow the step instructions to rename the layer, and then click Next Step.**

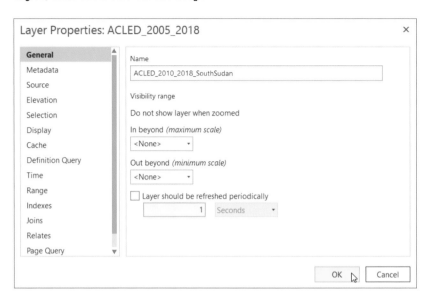

Step 3 of the task was configured to execute automatically. Your map zooms to the extent of the feature points.

8. **Click Finish to complete the task.**

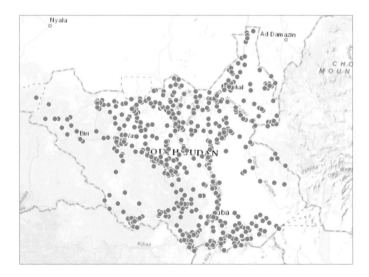

9. **Open the next task, Symbolize layer.**

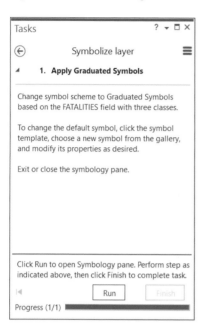

This task has only one step (you can tell because the Progress indicator says 1/1). The task author can give a task as few or as many steps as required.

10. **Now that you are familiar with the way the Tasks pane works, complete the Symbolize layer task on your own.** When you click Run, the Symbology pane opens.

11. Follow the task step instructions to configure symbols.

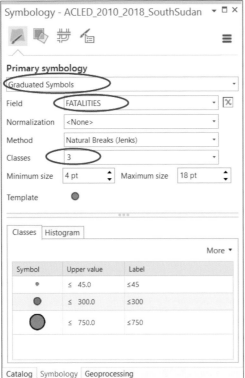

The next task calls a geoprocessing tool.

12. When you have finished the Symbolize layer task, open the Make selection task.
This task also has one step. The Select Layer By Attribute tool is embedded within the Tasks pane, with default parameters preselected. The selected features will be features with an EVENT_TYPE field value of "Violence against civilians." (Hint: point to the expression to see it in its entirety.)

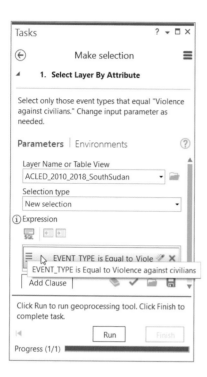

TIP *You can change geoprocessing tool parameters embedded in a task.*

13. **Click Run. When the process is complete, click Finish.**

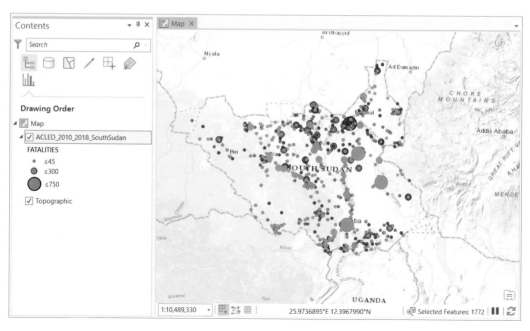

The map shows conflicts in South Sudan between 2010 and 2018, symbolized with graduated symbols, so that the increasing size corresponds to a greater number of fatalities. Conflicts that have been classified as violence against civilians are selected.

Author a task

If you want to create a brand-new task item, on the Insert tab, click Task > New Task Item. Here, you will add to an existing task item.

1. **In the Tasks pane, click the Options button > Edit In Designer.** This step opens the Task Designer. You will be working with both windows simultaneously, so you will want to read the upcoming instructions carefully; they will be explicit about whether you should be in the Tasks pane or the Task Designer.

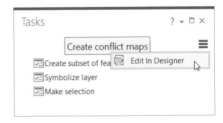

When the Task Designer is open, the Tasks pane reveals new buttons and tabs.

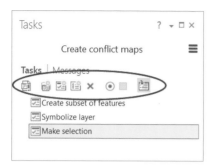

2. **In the Tasks pane, click the New Task button.** A new task is now visible in the Tasks pane, and the Task Designer is updated to reflect the new task. Clicking any task in the Tasks pane will bring up the same task in the Task Designer.

    TIP *When you have the Task Designer open, you can reorder tasks by clicking and dragging in the Tasks pane.*

3. **In the Task Designer, name the new task** Generate summary statistics. **For description, enter:** Summarize the fatalities for the selected set of features.

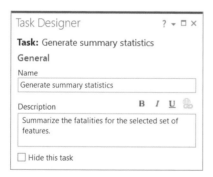

This task item will have one step. To populate the step, you will record your actions.

4. **In the Tasks pane, click the Record Commands button.** Your ArcGIS Pro interactions are now being recorded and will be translated into an automated step when you stop recording.

TIP *If you want to enter a step manually, without using the Record Commands option, click the New Step button.*

5. **On the ribbon, click the Analysis tab. In the Analysis Gallery, find and click the Summary Statistics tool. (Hint: you can also open ArcToolbox and search by tool name.)**

6. **Do not fill out the tool. Simply close the Geoprocessing pane (or keep it open if that is your preference).**

7. **In the Tasks pane, click the Stop Recording button.**

8. **Back in the Task Designer, for Run/ Proceed Instructions, enter:** Click Run to run geoprocessing tool. Click Finish to complete task. **Maintain the default step behavior (Manual).**

Task Designer   ? ▾ ☐ ✕

Step: Summary Statistics

General | Actions | Views | Contents

Name

Summary Statistics

Tooltip

Instructions   B *I* U 🔗

Run/Proceed Instructions   **B** *I* U 🔗

Click Run to run geoprocessing tool. Click Finish to complete task.

83 characters left

Step Behavior

◉ **Manual** User runs and user proceeds ⓘ

○ **Auto Run** Step runs and user proceeds ⓘ

○ **Auto Proceed** User runs and step proceeds ⓘ

○ **Automatic** Step runs and step proceeds ⓘ

☐ **Hidden** User will not see the step

☐ **Optional** User can skip this step

9. **In the Task Designer, click the Actions tab. Point to Summary Statistics, and click the Edit tool. Select the parameters listed as follows, and then click Done.**
   - **Input Table: ACLED_2010_2018_SouthSudan**
   - **Output Table: Conflict.gdb**ACLED_SouthSudan_Statistics
   - **Field: FATALITIES**
   - **Statistic Type: Sum**

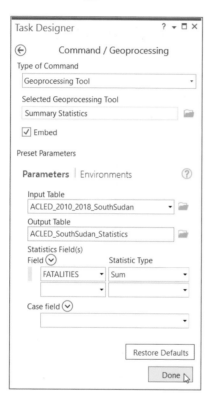

10. **Close the Task Designer.**

11. **Run the Generate summary statistics task, and then click Finish to complete the task item.**

12. **Open the new statistics table that has been added to the map.**

According to ACLED statistics, how many fatalities resulted from violent conflicts against South Sudanese civilians between 2010 and 2018?

TIP *To see the new table in the Catalog pane, right-click Conflict.gdb, and click Refresh.*

ON YOUR OWN

Repeat the entire task for another African country of your choice.

13. **Close the table and the Tasks pane, clear the selection, and save the project.**

ON YOUR OWN

In the Catalog pane, under Tasks, right-click Create conflict maps, and click Export to File > Save As. Name the task file as you like, and save it to your Conflict folder. You can now share the file with a colleague by attaching it to an email or putting it on a network. Saving the task file is a good way to share a workflow when you do not want to share an entire ArcGIS Pro project.

Tasks provide the benefit of a built-in, how-to tutorial without needing separate documents to explain workflows. Whether they are commonly performed tasks or more obscure workflows that are only performed occasionally, storing the sequence of steps in a task item is a good way to ensure compliance and consistency.

For more information on tasks, in ArcGIS Pro Help go to Workflows > Tasks.

You can continue to the next exercise, or close ArcGIS Pro and come back to it later.

EXERCISE 5B

Create a geoprocessing model

Estimated time to complete: 25 minutes

ModelBuilder, packaged within ArcGIS Pro, is a design environment for creating spatial analysis workflow diagrams. The diagrams you make are called *models*, and they have many advantages. A *model* allows you to string multiple geoprocessing tools together and run them automatically with the click of a button. The output of one process becomes the input for the next process, and so on.

In this exercise, you will execute a workflow like the one you did in exercise 5a; however, instead of using tasks, you will use ModelBuilder to build, save, and automate the workflow.

Exercise workflow

- Create a new ACLED layer with a limited feature definition.
- Build a model to do the following:
  - Apply the symbology from one layer to another.
  - Select only those conflicts classified as "violence against civilians."
  - Summarize the number of fatalities within the selected set of features.

Define the data

Before you jump into ModelBuilder, you will configure a new feature layer to serve as the model's first input dataset.

1. **In ArcGIS Pro, continue working with the Conflict project that you modified in exercise 5a (\GTKAGPro\Conflict\Conflict.aprx).** The project should have a layer

named ACLED_2010_2018_SouthSudan, which displays conflicts that occurred in Sudan between 2010 and 2018. Conflict locations are symbolized using graduated symbols so that larger symbols show greater fatalities. The Contents pane also has a statistics table.

TIP *If you did not successfully complete exercise 5a, add the layer file ACLED_2010_2018_SouthSudan.lyrx from the project folder.*

TIP *Recall that layer files are not actual data; they point to a source dataset (in this case, the ACLED_2005_2018 geodatabase feature class), but they provide layer properties such as definition queries, selections, or symbology schemes. You do not need the statistics table.*

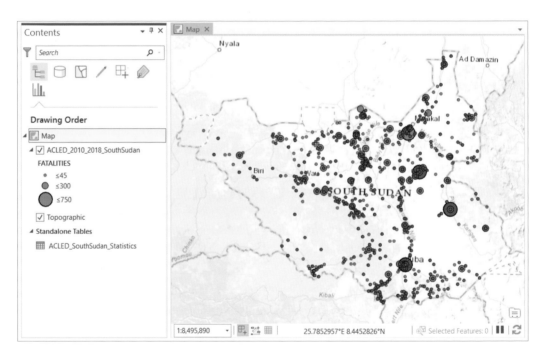

2. **From Conflict.gdb, add the _2005_2018 feature class to your map, and zoom to the layer.** This complete layer will cover the Sudan layer.

3. **Open the properties of the new layer, and click the Definition Query tab.**

4. **Build two definition query clauses:**
   - **COUNTRY is Equal to Rwanda**
   - **And YEAR is Greater Than or Equal to 2010**

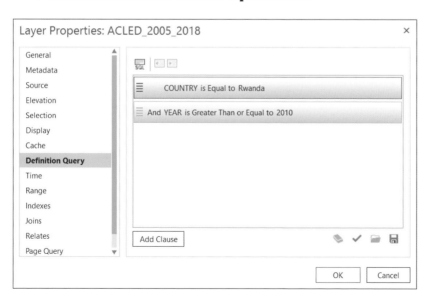

REMIND ME HOW

- *Click Add Clause.*
- *In the first drop-down box, scroll down and select COUNTRY. Maintain the default selection (is Equal to) for the second drop-down box. For the third drop-down box, select Rwanda. (If you do not see Rwanda at first, click More.)*
- *Click Add to finalize the first query.*
- *Again, click Add Clause. Configure the next four drop-down boxes to build the next query statement: And YEAR is Greater Than or Equal to 2010*
- *Click Add.*

5. **Click the General tab. Change the layer name to** ACLED_2010_2018_Rwanda, **and then click OK.**

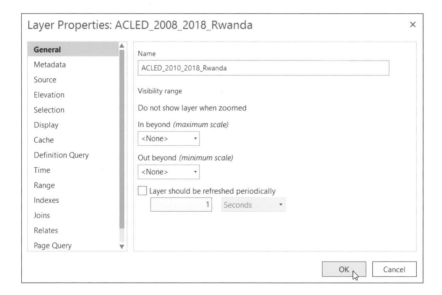

6. **Zoom to the extent of the new layer.**

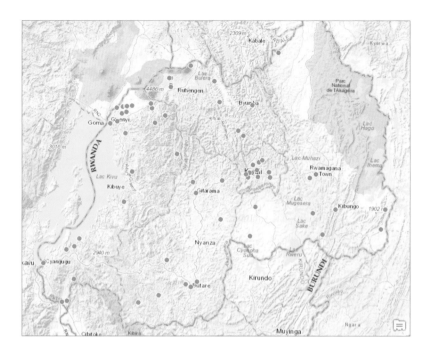

7. **Symbolize ACLED_2010_2018_Rwanda using graduated symbols based on the FATALITIES field. Use the default classification method but change the number of classes to 3, and modify the symbol template to your liking.**

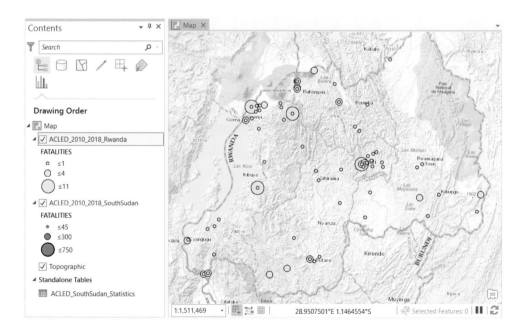

REMIND ME HOW

- *To open the Symbology pane, with the new layer selected in the Contents pane, click the Symbology button on the Appearance tab.*
- *In the Symbology pane, select Graduated Symbols from the Symbology drop-down list. Change the Field option to FATALITIES. Maintain the default classification options. To change the appearance of the points, click the symbol template. Choose a new symbol from the Gallery. Modify the properties as desired. Close the Symbology pane if you want.*

Add operations to ModelBuilder

Next, you will create a new geoprocessing model that will reside in your project's toolbox.

1. **On the Analysis tab in the Geoprocessing group, click the ModelBuilder button.** This step opens the ModelBuilder window. By default, it is docked in the map view area, on a new tab. You can undock the window if you prefer.

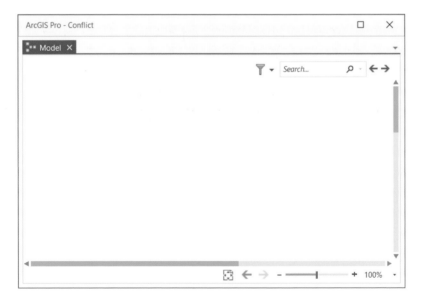

A ModelBuilder tab is now available on the ribbon.

TIP *Alternatively, you can right-click a toolbox in the Catalog pane, and then click New > Model.*

2. **In the Catalog pane, expand Toolboxes > Conflict.** Notice that the Conflict project toolbox contains a new empty model. (This model was automatically created when you opened the ModelBuilder window.) The toolbox also contains a script tool, which you will use in exercise 5c.

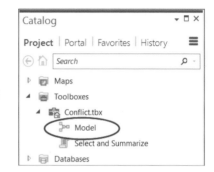

3. **On the ModelBuilder tab, in the Insert group, click the Tools button to open the Geoprocessing pane, if necessary.**

4. **In the Geoprocessing pane Search box, type** select. **From the Search Results window, drag the Select Layer By Attribute tool to the empty ModelBuilder window.**

5. **In the same way, search for** summary, **and drag the Summary Statistics tool to the ModelBuilder window.**

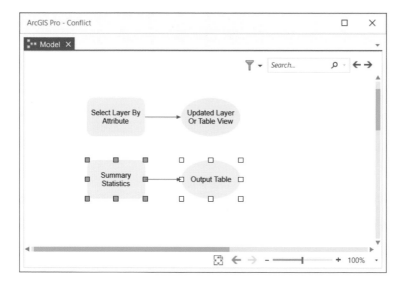

You have two empty tools in your model. (If your ModelBuilder window is floating, resize it as necessary.) Next, you will define the tool parameters.

Fill out the tool parameters

1. **In the model, double-click the Select Layer By Attribute tool.**

2. **For the first parameter, select ACLED_2010_2018_Rwanda. Maintain the default selection type (New selection). For Expression, click Add Clause. Use the drop-down boxes to build the following expression:**
   `EVENT_TYPE is Equal to Violence against civilians.`

3. **Click Add, and then click OK.**

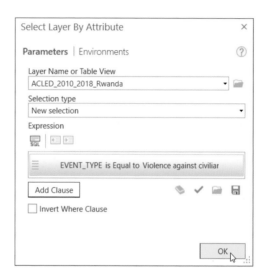

The tool in the model now shows an input layer and is shaded, which indicates that it is ready to run.

TIP *To see the entire input and output layers in the model, click an element and resize it by dragging the selection anchors. If the ModelBuilder window is floating, resize it as necessary.*

4. **In the model, double-click the Summary Statistics tool.**

5. **For Input Table, select ACLED_2010_2018_Rwanda:(2).** You are choosing intermediate data; that is, the output of the first operation becomes the input of the second operation. Do not worry if the text box still shows Rwanda:1. If you select Rwanda:(2), it will connect properly.

6. **For Output Table, maintain the default location, but replace the default name with** ACLED_ Rwanda_Statistics. **For Field, select FATALITIES. For Statistic Type, select Sum. Leave the Case field parameter blank. Click OK.**

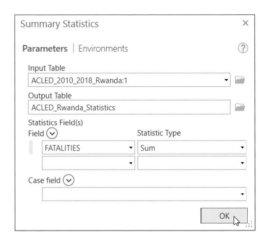

ModelBuilder automatically connects this operation to the output of the selection operation. The statistics will be calculated only for selected features.

7. **On the ModelBuilder ribbon, click the Auto Layout button.** **Resize the elements or the ModelBuilder window as needed, and then click the Fit To Window button.**

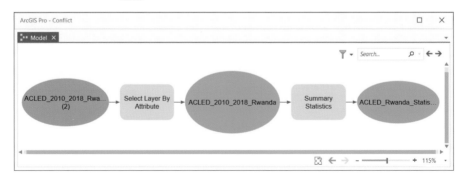

8. **On the ModelBuilder tab, in the Model group, click the Save button to save the model.**

    TIP *Alternatively, click Save As, and give the model a unique name, such as* ACLED_select_stats.

    You can run tools in a model one by one (simply right-click the tool, and click Run), or run connected tools automatically, which is what you will do next.

Run the model

1. **On the ModelBuilder tab, in the Run group, click Run.** A progress window appears to indicate the model's status. You may also notice that each tool in the model turns red as it is processing.

2. **When the model has completed all of the processes, close the progress window.**

3. **Observe the map.** In the Rwanda layer, conflicts classified as "violence against civilians" are selected.

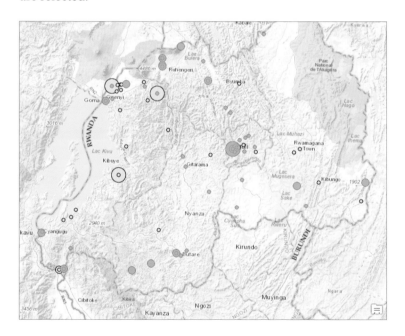

4. **In the Catalog pane, open Conflict.gdb, and notice that it contains a new table.**

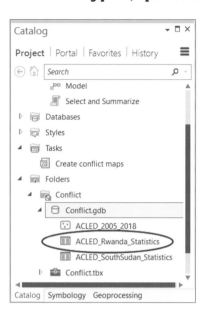

TIP *If you do not see the table, right-click*
the geodatabase, and click Refresh.

5. **Drag the table to the map, and open it to review the data.**

How many fatalities resulted from conflicts classified as "violence against civilians" in Rwanda from 2010 to 2018?

This model was a simple one. Models can be as simple or as complex as needed. The value of a model is that you can easily modify the input dataset and any tool parameters as necessary; you can save the model with a new name if desired, and run the model on a new dataset. You can tweak, refine, and add to the model whenever you want, without having to start from scratch. You can also set model parameters so that the model can be run from a geoprocessing tool dialog box, which you will do next.

Convert a model to a geoprocessing tool

You can turn a model into a stand-alone geoprocessing tool that you can easily share and then repeat the process, including changing the input, output, and variables. Here is how.

1. **In ModelBuilder, right-click the input element (the blue oval), and click Parameter.**

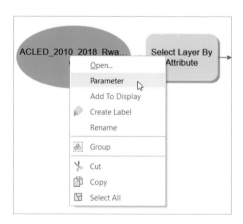

2. **Right-click the Select Layer By Attribute tool element (the first yellow rectangle), and click Create Variable > From Parameter > Expression.**

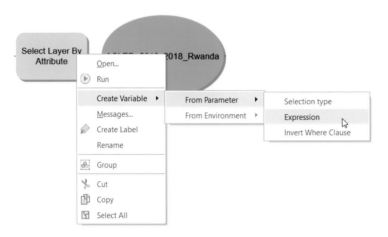

3. **Right-click the Summary Statistics tool element, and click Create Variable > From Parameter > Statistics Field(s).**

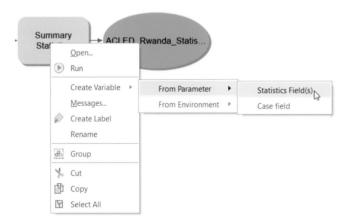

4. **Set the final output as a parameter. Also set both variables as parameters.**

5. **Move the variable elements as you want, or click the Auto Layout button.**

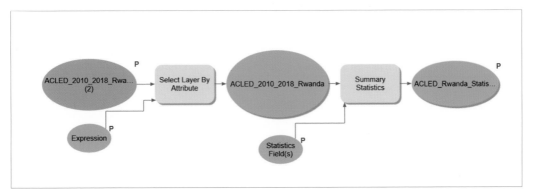

6. **Save the model, and close ModelBuilder.**

7. **In the Catalog pane, find the model tool in the Conflict toolbox, and double-click it to open it.** Your model looks like a regular geoprocessing tool.

8. **Change the output variable to ACLED_Rwanda_Statistics2, edit the Event type to Riots/Protests, and then run the tool.** The new statistics table is automatically added to the map.

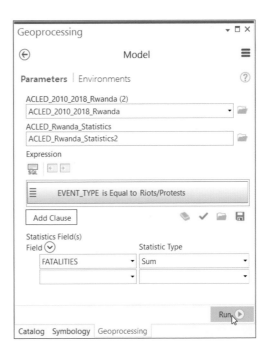

*How many events classified as riots/protests occurred in
Rwanda between 2010 and 2018? How many fatalities resulted
from these events?*

You can also repeat the first seven steps of this exercise ("Define the data"), choosing a different country to display, and then choose that new layer as your model input, varying the selection and statistics field as desired.

9. **Clear the selection, close the model and any open tables, and save the project.** You will continue using the model in exercise 5c.

You will use ModelBuilder again in chapter 9.

EXERCISE 5C

Run a Python command and script tool

Estimated time to complete: 25 minutes

You now have experience executing geoprocessing functions in two ways: using stand-alone tools and combining them in the ModelBuilder window.

You can also run geoprocessing tools, along with other functions, using Python. *Python* is a programming language that, in the GIS context, is used to *script* geoprocessing workflows and build custom geoprocessing tools.

You can type Python commands from the Python window (*command line*) in ArcGIS Pro. You can also install Python for ArcGIS Pro to run Python commands and scripts outside the application.

ArcPy is the name of the Python site package that enables you to run ArcGIS commands from Python.

If the idea of writing code does not appeal to you, do not worry. A full synopsis of Python and coding techniques is far beyond the scope of this book. Think of this exercise as a first look at the Python window in ArcGIS Pro, something you may want to explore in depth later.

Even if you are well-schooled in coding, this exercise will be helpful to see how you can incorporate Python into ArcGIS Pro.

You will follow a now familiar workflow, like the workflow in exercises 5a and 5b. But this time, you will use Python coding and a script tool to execute the processes.

1
2
3
4
5
6
7
8
9
10

Exercise workflow

- *Create a new ACLED layer using a limited feature definition and meaningful symbology.*
- *Use a geoprocessing tool and then Python code to do the following:*
  - *Create a layer file (.lyrx) from the new map layer.*
- *Use a custom script tool to do the following:*
  - *Select only those conflicts classified as "violence against civilians."*
  - *Summarize the number of fatalities within the selected set of features.*

Define the data

Before jumping into what might be your first experience in writing code in ArcGIS software, you will set up a new map layer.

1. **Continue working in Conflict.aprx.** It should contain two ACLED layers (ACLED_2003_2013_Sudan and ACLED_2003_2013_Rwanda), both symbolized using graduated point symbols based on the FATALITIES field.

   TIP *If you did not successfully complete exercises 5a and 5b, add the layer files ACLED_2010_2018_SouthSudan.lyrx and ACLED_2010_2018_Rwanda.lyrx from the \GTKAGPro\Conflict folder.*

2. **Add another copy of the ACLED_2005_2018 feature class to the map. Limit the layer's definition to the country of Nigeria, displaying conflicts between 2010 and 2018.**

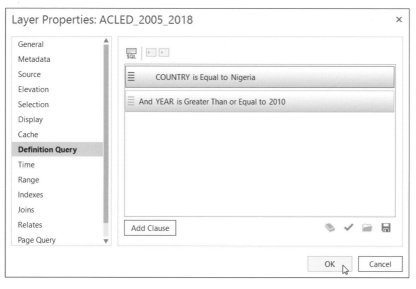

3. **Rename the layer** ACLED_2010_2018_Nigeria, **and zoom to the defined layer.**

4. **Symbolize ACLED_2010_2018_Nigeria using graduated symbols based on the FATALITIES field. Use the default classification method, change the number of classes to 3, and change the symbols to whatever you prefer.**

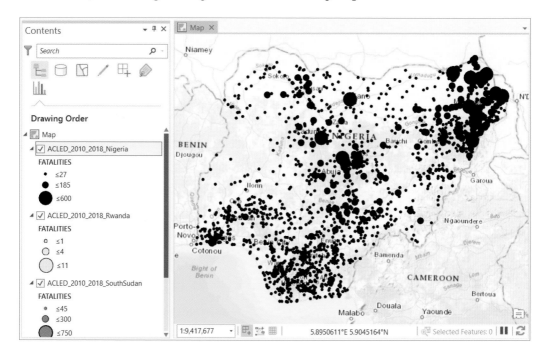

Execute a command using Python

Next, you will create a layer file to store the definition query, name, and symbology scheme of this layer. Then it can be used in other maps or projects without having to repeat the preceding steps. Layer files also save label characteristics and custom layer pop-up windows, if applicable. You will first execute this command using the Save To Layer File geoprocessing tool, then again using copied Python code, and a third time by entering the code yourself.

1. **On the Analysis tab, click the Tools button to open the Geoprocessing pane.**

2. **Enter the search word** save, **and then open the Save To Layer File tool.** The tool parameters require an input layer, an output file name, and an optional setting to save the file using a relative path. (Using this option means that if you move the layer file to another location, its path to the source data is updated accordingly.)

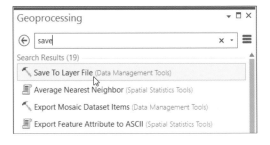

3. **Set the following tool parameters, and then click Run:**
   - **Input Layer: ACLED_2010_2018_Nigeria.**
   - **Output Layer: \GTKAGPro\ Conflict**\ACLED_2010_2018_ Nigeria.lyrx.
   - **Select the check box to store relative paths.**

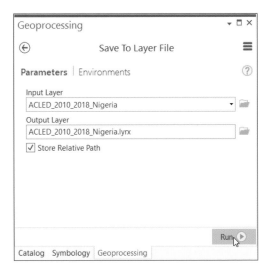

4. **When the tool is finished running, take a moment to save the project.**

5. **Right-click the status bar at the bottom of the Geoprocessing pane, and click Copy Python command.**

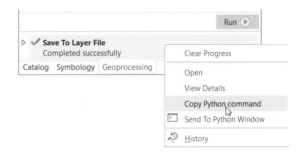

TIP If you already closed the Geoprocessing pane, you can alternatively copy the Python command from the Geoprocessing History in the Catalog pane.

6. **On the View menu, click the Python button.** This step opens the Python pane. Like other panes, it can be docked in one of several locations or free-floating.

7. **At the bottom of the Python pane, right-click in the code entry box, and click Paste.**

TIP All geoprocessing tools are accessible from ArcPy. If you have a workflow that uses 10 geoprocessing tools, you can write 10 lines of code, with one line calling each tool.

8. **Click at the end of the copied code to see a code parameter description pop-up message.**

Just like the geoprocessing tool, the Python code parameters expect you to enter an input layer and output layer. Optionally, you can enter a parameter to set the relative path option and choose which version of the data to use when multiple versions exist. Optional parameters are marked with curly brackets {like this} in the description pop-up message.

If you run the code as is, the operation will fail because there is already a layer file in the Conflict folder named ACLED_2010__2018_Nigeria.lyrx—the one you just created from the Geoprocessing pane.

9. **In the Python window, change the output file name to ACLED2010_2018_ Nigeria2.lyrx.**

```
arcpy.management.SaveToLayerFile("ACLED_2010_2018_Nigeria", r"C:\EsriPress
\GTKAGPro\Conflict\ACLED_2010_2018_Nigeria2.lyrx", "RELATIVE", "CURRENT")
```

10. **Again, click at the end of the code, and then press Enter to run the command.**
    When processing is complete, a second Nigeria layer file is added to the Conflict folder.

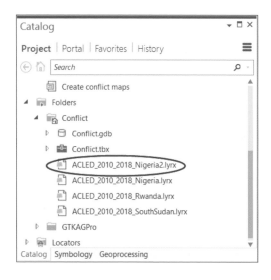

Now you will run the Save To Layer File tool a third time, but this time you will enter the code manually.

TIP *To clear the history in the Python window, right-click the upper portion of the window, and click Clear Transcript.*

ON YOUR OWN

The following steps are optional. At least read through them so you understand that there are additional steps required if you are working outside ArcGIS or your desired geoprocessing workspace.

- *In the code entry portion of the Python window, type the following:*
  ```
  import arcpy
  ```
- *Press Enter. The line of code is added to the upper portion of the Python window.*

This line is a matter of protocol. Within ArcGIS applications, it is not required. But if you want to run Python commands outside ArcGIS, it is required, so adding the line is a good habit to get into.

Next, you will set the environment to ensure that the operation's output will be stored where you want it. Again, this step is optional, because you are already in the workspace you are calling, but if you are working in a different project or environment, it is required.

- *Type the following code, and then press Enter:*
  ```
  arcpy.env.workspace = 'C:/EsriPress/GTKAGPro/Conflict'
  ```

Single quotation marks and double quotation marks are treated the same; for example, you can enter 'name' or "name." Also, paths are not case sensitive. So you can also type it this way:
```
arcpy.env.workspace = "c:/esripress/gtkagpro/conflict"
```

11. **Begin to type the following line:**
    ```
    arcpy.SaveToLayerFile
    ```

    Notice that the Python pane offers the correct entry when you begin typing.

12. **Click the auto-complete suggestion to call the tool.** The highlighted pair of parentheses after the tool and toolbox name is where you add your tool parameters.

13. **Click between the parentheses. Notice that you are offered a list of map layers as possible inputs. Click ACLED_2010_2018_Nigeria.**

TIP *You can also drag inputs from the Contents pane.*

14. **After the input dataset (following the quotation mark but inside the closing parenthesis), type a comma and then a single quotation mark. Notice that the closing quotation mark is provided for you. Type the output layer file name shown here, or choose a different name.**
    ACLED_2010_2018_Nigeria3.lyrx

15. **After the output quotation mark, type another comma, and then from the options box choose either 'RELATIVE' or 'ABSOLUTE'.**

16. **Type another comma, and then choose the only available option ('CURRENT').**

The command is ready to run.

17. **Click after the closing parenthesis, and then press Enter.**

18. **When the process is complete, find the new layer file in the project folder.**

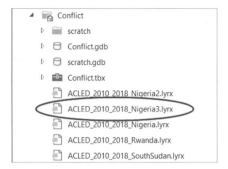

19. **Close the Python window.** You learned how to use Python by both copying existing code and typing it manually. You can delete the redundant layer files if you want.

Use a custom script tool

Script tools look like any ArcGIS geoprocessing tool, except that they are based on a Python script. You can use Python to create a custom geoprocessing tool. Similar to creating a geo-processing tool based on a model, the tool author specifies the tool parameters that the end user can modify.

1. **In the Conflict toolbox, notice the Select and Summarize custom script tool. To view the code behind the tool, right-click the tool, and click Edit.**

```
# Select and Summarize
# --------------------------------------------------------------------------
# Created on: 2018-08-10
#
# Description: The user can select an input file to symbolize and an output file to store the summary res
#
# --------------------------------------------------------------------------

# Import arcpy module
import arcpy
from arcpy import env

# Set the environment for output file location
env.workspace = ("c:\EsriPress\GTKAGPro\Conflict\Conflict.gdb")

# Prompt the user for the name of the input layer
ACLED_2010_2018_InputFile = arcpy.GetParameterAsText(0)
# Prompt the user for the name of the output file
ACLED_2010_2018_Output_Stats = arcpy.GetParameterAsText(1)

# Process: Select Layer By Attribute
arcpy.SelectLayerByAttribute_management(ACLED_2010_2018_InputFile, "NEW_SELECTION", "EVENT_TYPE = 'Violen

# Process: Summary Statistics
arcpy.Statistics_analysis(ACLED_2010_2018_InputFile, ACLED_2010_2018_Output_Stats, "FATALITIES SUM", "")
```

> **TIP** *Anything preceded by a hashtag (#) is a comment and is ignored by Python; it is for the user's benefit.*

The code includes the tool name, creation date, and description, followed by the processing instructions: the call to import arcpy, environment settings, geoprocessing tool parameters, and the geoprocessing tools executed by the script.

Which geoprocessing tools are combined in this script?

> **TIP** *If your data is not stored in C:\EsriPress\GTKAGPro, you should manually edit the "env.workspace = " line. Then click File > Save.*

2. **Close the window. Now right-click the Select and Summarize tool, and click Open.**
    You saw the back end of the tool a moment ago—the code behind it. This view is the front end, and it looks just like any other geoprocessing tool. But instead of performing one process, this tool performs two, and it can be more than two. Script tools are especially good for automated workflows and batch processing.

3. **For Select Input Layer, select ACLED_2010_2018_Nigeria.** The selection will be made in this layer. Recall that the second process, after the attribute selection, is to create a table of summary statistics based on the total fatalities within the selected set (violence against civilians).

4. **For the output file name, enter** ACLED_Nigeria_Statistics. The tool is already programmed to store the table in the Conflict.gdb workspace.

5. **Now run the tool.** Notice that the Nigeria conflicts that are characterized as "violence against civilians" have been selected, as indicated by the script tool code. The tool also created a table that summarizes the Fatalities value of the selected set.

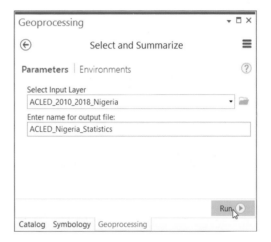

6. **Find the new table in the Catalog pane (\Folders\Conflict\Conflict.gdb). (Refresh the geodatabase if necessary.) Drag the table to the map, and open it.**

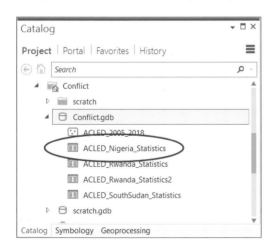

How many fatalities resulted from conflicts classified as "violence against civilians" in Nigeria from 2010 to 2018?

This introduction to Python is only cursory. You began and did some programming, which is great, even if you did not fully understand the underlying concepts of Python. Many resources are available that can give you more thorough lessons in using Python with ArcGIS Pro. For starters, explore the ArcGIS Pro ArcPy Reference at http://pro.arcgis.com.

If you choose to share your saved project, you are also sharing the task, model, and script tool that you created in this chapter.

7. **Clear the selection, and save the project.**

Package the project

To share your project, including the task, model, and script tool you created, you must create a project package.

1. **Click the Share tab, and then click the Project button.**

2. **In the Package Project pane, maintain the option to Upload package to Online account.**

3. **Maintain the default name. Under Summary, enter a sentence or two that explains the project.**

4. **Under Tags, enter some words that you think capture the key points about the project (such as** conflict, Africa, **and** ACLED**).**

5. **Save the project to your root folder, and choose to share it with your organization or selected groups. Click Package.**

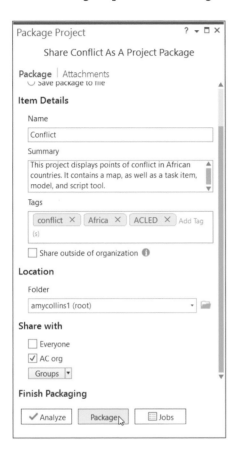

ON YOUR OWN

Log in to ArcGIS Online (www.arcgis.com). Under Content, find the project.

You are finished with this chapter. If you are not continuing to chapter 6, you can close ArcGIS Pro.

Summary

You now have some experience using ArcGIS Pro tasks, ModelBuilder models, and Python scripting to automate GIS workflows. As you continue using GIS in your own profession, you may find that one environment is best suited for solving a problem, or you may find that you can incorporate all three environments. Now that you have taken a first look at each method, you should feel comfortable doing more investigation and learning on your own.

Glossary terms

task

definition query

query expression

ModelBuilder

model

Python

script

command line

CHAPTER 6
Collaborative mapping

Exercise objectives

6a: Prepare a database for data collection

- Set up a domain
- Set up feature class fields
- Symbolize features
- Publish features

6b: Prepare a map for data collection

- Create a map
- Set up additional data display options
- Share the map and enable editing

6c: Collect data using Collector for ArcGIS

- Open Collector for ArcGIS
- Collect a tree location and enter the data
- View the map in ArcGIS Pro

Digital data collection technology is quickly evolving to allow you to simultaneously collect data and collaborate on map projects. The idea is that a group of users can work together to gather and record information—also known as *crowdsourcing*. Collaborative mapping can be used in many ways. It is often seen in the form of social mapping—for example, at the site of a weather-related disaster such as a flood or storm. People send updates, photos, videos, and other details to various social media channels. Although some of the data can be viewed geographically in real time, much of this data is analyzed at a later time to identify trends and patterns. Data collection in the realm of GIS is generally more specific and incorporates a more structured approach of capturing the information, such as creating a map of all of the homes for sale in an area, including the details about each home (such as sale price, number of bedrooms, and square footage).

Users can collaborate to collect data and create maps in many ways. At the most basic level, users can simply collect information using pen and paper. Although this type of collaboration can work, it may not be as effective if more users are added to collaborate or if the type of data being recorded is more detailed. Even if the single piece of paper is a very large one, there is a physical limit to how many people can work at the same time on that piece of paper. The ArcGIS platform allows groups to collaborate on maps electronically, which expands the number of potential users.

Scenario: you are part of a community tree inventory program that is doing a field-based assessment to generate data of trees in your area. You are tasked with setting up a collection system that you can share with others in the program. This type of field work is a bottom-up approach using on-the-ground data collection as opposed to a top-down approach using aerial or satellite imagery. On-the-ground methods can provide a good approximation of tree characteristics, such as the number of trees in an area, the location of trees measured by GPS, tree species, tree size, and tree health, including risk of disease. Although a bottom-up approach requires careful planning, management, and considerable cost and effort, it can provide detailed information that a top-down analysis cannot.

In the following exercises, you will learn how to complete a tree inventory database in ArcGIS Pro that consists of information including tree location, diameter, height, condition, and hazards. You will publish the data to your ArcGIS Online organizational account so that it can be used to create a map that can be accessed and used with Collector for ArcGIS, a mobile app for your smartphone or tablet.

> **TIP** *Completing these exercises requires an ArcGIS Online organizational account. You can sign up for an ArcGIS trial by going to the Esri Press book resource page (esri.com/gtkagpro2).*

EXERCISE 6A

Prepare a database for data collection

Estimated time to complete: 25 minutes

In this exercise, you will complete a tree inventory geodatabase in ArcGIS Pro by setting up a domain, creating a feature class, and adding additional fields. The tree inventory database will be used to store information collected using the Collector for ArcGIS mobile app. Once you configure the information model, you will build a map in ArcMap and publish a feature service to your *organization*, a shared online workspace that is tied to your software license.

Exercise workflow

- *View existing domains and set up a new domain to collect tree status data.*
- *View existing fields and set up a new field based on the status domain.*
- *Symbolize points of the Trees feature class.*
- *Publish the Trees feature class to your ArcGIS Online organization.*

Set up a domain

Domains provide a way for you to constrain input information by limiting the choice of values for a particular field. Coded values limit what can be chosen when collecting data, helping to standardize data collection from the field. Using domains helps maintain data integrity and does not allow other values to be entered during data collection.

1. **In ArcGIS Pro, open TreeInventory.aprx, found in your \EsriPress\GTKAGPro\ TreeInventory folder.**

2. **In the Catalog pane, click Project, and expand Databases. Right-click TreeInventory, and go to Design > Domains**

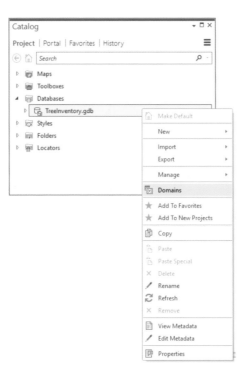

The Domains: TreeInventory pane opens. The two existing domains are Hazard and LandUse. A brief description explains what each domain is used for. (Your screen may look different, depending on whether you have the Domains: TreeInventory pane docked or floating.)

3. **Click the Hazard and LandUse domain rows to view the coded values they contain.**

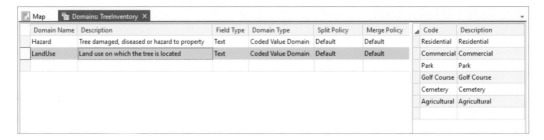

The coded values populate a list of choices that you can select from when collecting data.

4. **Click the empty row below the LandUse row to select it. Click the Domain Name field, and type** Status. **Press the Tab key or click the Description field, and type** Tree status.

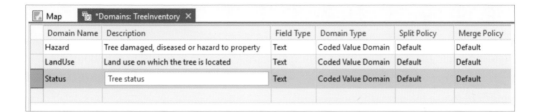

5. **On the right side of the Domains: TreeInventory pane, click the first empty row to highlight it. In the Code field, type** Planted **for the first valid code. Press the Tab key or click the Value field, type** Planted, **and then press Enter.**

TIP *You can use your own abbreviated codes in the Code field. Abbreviated codes are useful when coded values are long or contain multiple words.*

6. **Add three more coded values to the Status domain:**
   - Ingrowth
   - Unknown
   - Dead

A planted tree is a tree that was intentionally planted. An ingrowth tree is a tree that was self-seeded. An unknown tree is a tree whose status as planted or ingrowth cannot be determined. A dead tree is a tree that is no longer alive. When collecting data, you can select from only these choices to set a tree's status.

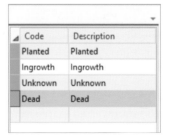

7. **On the Domains tab, click Save.**

8. **Close the Domains: TreeInventory pane.** Next, you will set up fields in the Trees feature class to use the newly created domain.

Set up feature class fields

Fields in a feature class store information and are a key part of any data collection model. You will explore the feature class in the TreeInventory database to see how it is set up and create a new field to use the domain you created.

1. **In the Contents pane, click Project, and expand the TreeInventory database. Right-click Trees, and go to Design > Fields.**

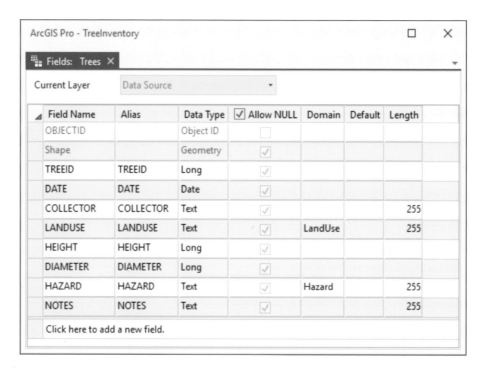

This step opens the Fields: Trees pane. The Trees point feature class contains fields that are commonly used in tree inventory data collection. The following is a short description of the existing fields:

- TREEID—a tree identification number
- COLLECTOR—person who collected the data

- HEIGHT—tree height
- HAZARD—identifiable tree hazards
- NOTES—other notes about the tree
- DIAMETER—tree diameter
- LANDUSE—land use on which the tree is located
- DATE—when the data collection took place

Also look at the other columns for each attribute field. The *data type* determines what kind of data that the field can store; the *alias* can be used to refer to the attribute field by a different name; and the domain determines which existing domain the field can use. You will add a new tree status field so that it can be assigned the Status domain.

2. **Below all of the fields, click "Click here to add a new field" and type** STATUS **for Field Name. Under Alias, type** STATUS. **Under Data Type, select Text.**

ArcGIS Pro - TreeInventory							□ ×

Fields: Trees ×							▾

Current Layer	Data Source ▾

Field Name	Alias	Data Type	☑ Allow NULL	Domain	Default	Length	
Shape		Geometry	☑				▲
TREEID	TREEID	Long	☑				
DATE	DATE	Date	☑				
COLLECTOR	COLLECTOR	Text	☑			255	
LANDUSE	LANDUSE	Text	☑	LandUse		255	
HEIGHT	HEIGHT	Long	☑				
DIAMETER	DIAMETER	Long	☑				
HAZARD	HAZARD	Text	☑	Hazard		255	
NOTES	NOTES	Text	☑			255	
STATUS	STATUS	Text ▾	☑			255	
Click here to add a new field.							▼

3. **Under Domain, click the Domain field to highlight it, and then click it again to select Status from the drop-down list.**

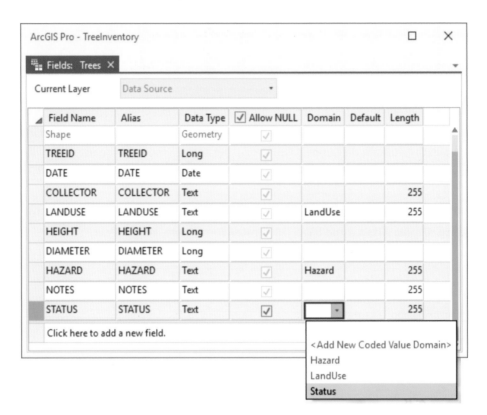

The tree Status domain selected here is the one you just set up so that you can select one when collecting data later.

4. **On the Fields tab, click Save.**

5. **Close the Fields: Trees pane.**

Symbolize features

You will add symbology to the Trees feature class based on the coded values of the Status attribute so that they are more distinguishable when collecting data.

1. **Add the Trees feature class to the current map.** Don't worry if you don't see anything—the feature class does not have any features yet. That is why no points were added to the map.

2. **In the Contents pane, right-click Trees, and click Symbology.**

3. **In the Symbology pane, click Primary Symbology, and select Unique Values from the drop-down list. For the Value field, select STATUS.**

4. **In the Classes tab, click the More down arrow, and go to Format all symbols.**

5. **Type park in the Search box, and press Enter. Make sure the drop-down list next to the Search box is set to All styles. Scroll to the Pushpins category, and click the medium-size Park symbol.** Project styles are styles that you add. Since you did not add any styles in this instance, you searched for the park style within all available styles included in ArcGIS Pro. All of the Status values are now assigned the medium-size green park symbol. Next, you will edit the color of the symbols individually to differentiate them.

    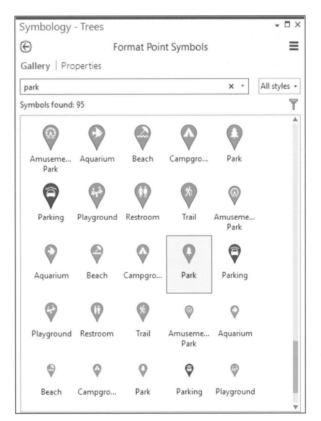

6. **Click the back button to return to the main Symbology pane.**

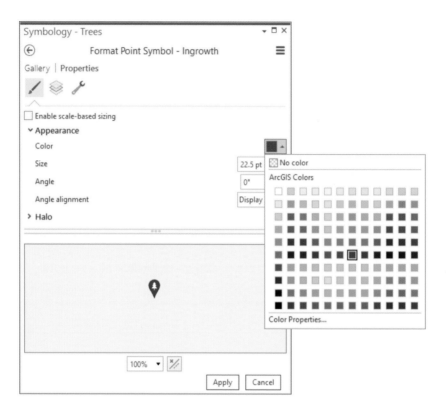

7. **In the Classes tab, click the Ingrowth symbol. Click Properties, and change the appearance color to a dark green (such as Fir Green). Then click Apply.**

8. **Repeat the same steps, and for the Unknown symbol, choose an orange color (Seville Orange). For the Dead symbol, choose a dark-red color (Tuscan Red).**

> **TIP** *To quickly switch between different symbol properties in the Symbology pane, click a symbol in the Contents pane to make it active, and then adjust its properties.*

9. **Click the More down arrow, and select the "Show all other values" check box to disable it.**

10. **Save the project.** Next, you will publish the layer to your ArcGIS Online organization.

Publish features

Before you can access your tree inventory data through the web or the Collector for ArcGIS app, you must publish it to your ArcGIS Online organization. Once published, the data will be available as a *web layer* that you can add to your maps.

TIP *You must be signed into ArcGIS Pro with your organizational account.*

1. **On the Share tab, in the Share As group, click Web Layer.**

2. **For Name, type** TreeInventory.

   TIP *Point to the blue information icons to learn about the difference between publishing features and publishing tiles.*

3. **For Description, type** A tree inventory layer for data collection.

4. **For Tags, type** tree, inventory, collector.

5. **At the bottom of the pane, click the Analyze tab.**

It is good practice to analyze the web layer before publishing to check for potential errors. Two warnings appear, with a summary of each warning. Selecting a warning provides more information about it. The "24041 Layer does not have a feature template set" warning occurs because you did not use a feature template that sets attributes by default. Do not worry though, because a default template is created for you upon publishing. The "24078 Layer's data source is not supported" warning is common when you have a basemap in your map. It will be dropped from the web layer upon publishing.

6. **Click Publish, and then click Jobs to see the progress.** The Trees feature class takes a few moments to process and upload. When a message confirms that it was successfully published, it is ready to view in ArcGIS Online.

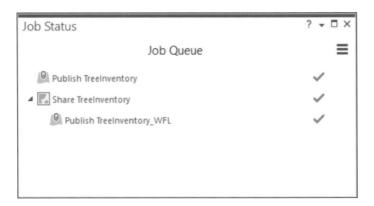

7. **Close the Share Web Layer pane.**

8. **Save the project.**

> **TIP** *The steps you took showed how to share a web layer, which will subsequently be added to a web map in ArcGIS Online where pop-up windows can be configured. This workflow is just one of several that you could use. You can also configure pop-up windows in ArcGIS Pro, and then share your map and layers as a web map.*

You have successfully set up the Tree feature class fields and published it to your ArcGIS Online organizational account. In the next exercise, the service will be leveraged as a layer in a map and used by your field work force to collect information in the field. You can continue to exercise 6b, or close ArcGIS Pro and come back to it later.

EXERCISE 6B

Prepare a map for data collection

Estimated time to complete: 45 minutes

In exercise 6a, you completed a tree inventory geodatabase and published it to your ArcGIS Online organization. In this exercise, you will sign in to your ArcGIS Online account and create a map and add the published tree inventory layer to it. This map will be used as the basis of collecting tree data in the field. You will also learn to configure and customize the data collection form directly through ArcGIS Online. When completed, you will share the map to make it available to others in the field who can access the Collector for ArcGIS app.

> **TIP** *You can complete this workflow using your organization. Notes are provided to help you. If you are not a member of an organization, create a trial organization to use.*

Exercise workflow

- *Create a map using the Tree Inventory feature layer.*
- *Set up additional data collection parameters.*
- *Share the map so that it can be used by others.*
- *Use the Collector for ArcGIS app to collect tree data in the field.*

Create a map

1. **Open a web browser, go to ArcGIS Online (https://www.arcgis.com), and sign in.**
   Sign in to the same ArcGIS Online organizational account that you used in exercise 6a.

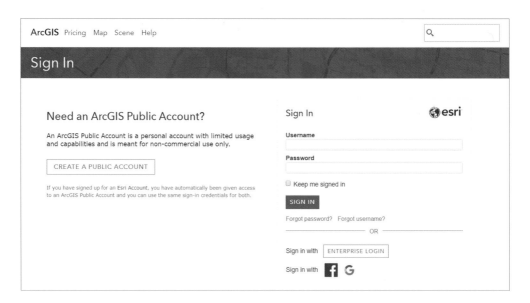

2. **Type your user name and password, and click Sign In.** Your organization's home page appears.

3. **On the main navigation bar, click Content.**

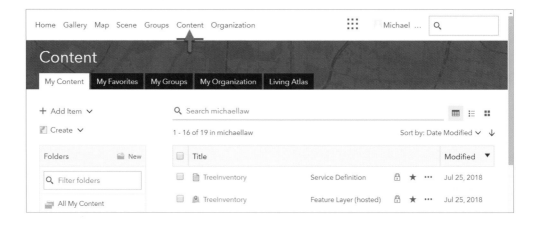

On the My Content tab, the TreeInventory feature layer that you published is listed (do not worry if your My Content page looks different from the graphic).

4. **If necessary, scroll or sort the columns to find the TreeInventory feature layer. In the TreeInventory (feature layer) row, click the More button, and from the drop-down list, select "Add to new map with full editing control."**

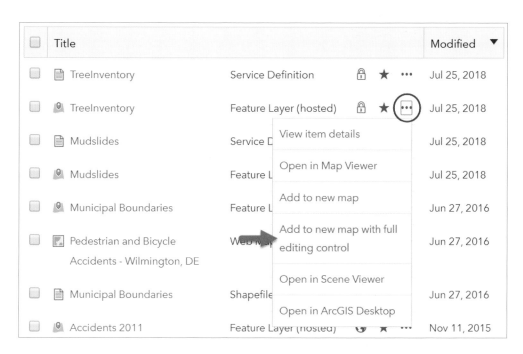

A new map named TreeInventory opens. It shows the default ArcGIS Topographic basemap. The map opens with a default extent. Do not worry if your map looks different from the graphic.

Because you added the TreeInventory feature layer to the map with editing enabled, you can immediately begin to edit the map. The Change Style pane opens by default. Click the When you close button to see the Contents pane, the TreeInventory feature layer is shown (by default as collapsed). You can click the TreeInventory layer name to expand it to see the symbols you set up in ArcGIS Pro (or click the Show Legend button).

5. **Pan or zoom to your geographic location. Alternatively, use the Search tool to find your address or place.** At a higher zoom level, you can add tree features to the map in better detail. Do not worry if your map or location does not match the exercise graphics.

6. **On the toolbar, click Edit. In the Add Features pane, click the Planted symbol. A ToolTip appears next to the mouse cursor. Click a location on the map to place the tree symbol.**

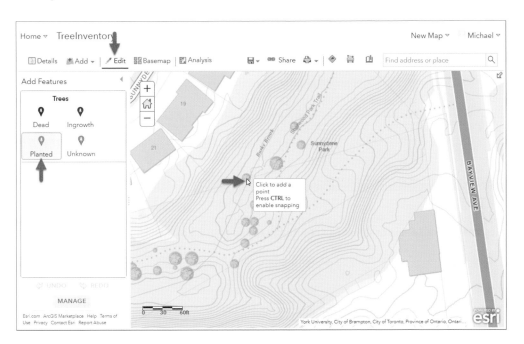

7. **Complete the tree inventory data collection form:**
   - **Type** 100 **in the TREEID field.**
   - **Click the DATE down arrow, and select today's date.**
   - **Enter your name in the COLLECTOR field.**
   - **Select a suitable LANDUSE type.**
   - **Include some values for the tree height and diameter (see the graphic as an example).**

8. **Click Close when you are finished completing the form.**

9. **Repeat and add several more tree features around your map.**

10. **Click the Save button, and click Save.**

The Save Map dialog box appears. This workflow provides an example of how you can create data from sources such as aerial photographs. You can use data collection in the field at the same site to ground truth what is being collected.

11. **Complete the fields in the Save Map dialog box as shown or as appropriate. Then click Save Map.**

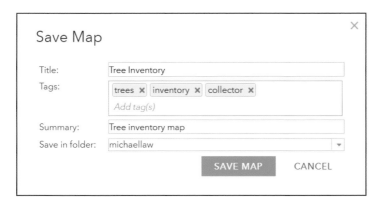

Set up additional data display options

1. **Click a tree feature to view its pop-up window.**

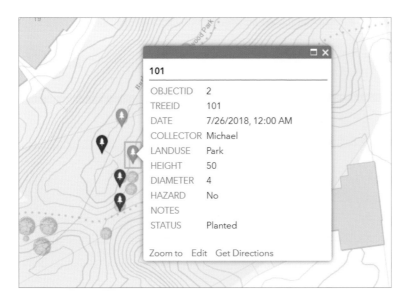

The tree inventory details you entered are displayed in the pop-up window. Because you are the data and map owner, you can set up additional data display options before finalizing the tree settings for the field.

2. **In the Contents pane, click the TreeInventory More Options button, and click Configure Pop-up. (If the TreeInventory layer is not visible, click Detailsfirst.)**

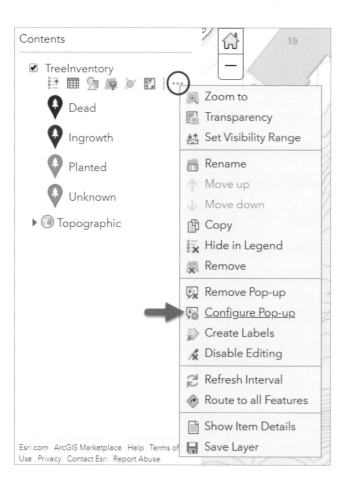

3. **Change Pop-up Title to**
    Community Tree Inventory. Field
    names in curly brackets dynamically
    pull information from the database
    to populate fields. You can configure
    how the field name labels look.

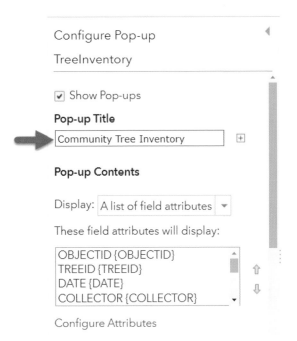

Configure Pop-up

TreeInventory

☑ Show Pop-ups

Pop-up Title

Community Tree Inventory   ⊞

Pop-up Contents

Display: A list of field attributes ▾

These field attributes will display:

OBJECTID {OBJECTID}
TREEID {TREEID}
DATE {DATE}
COLLECTOR {COLLECTOR}

Configure Attributes

4. **Click the Configure Attributes link.**

5. **In the Field Alias column, click HEIGHT, and type** (FT) **after it. Repeat for the**
    DIAMETER field alias.

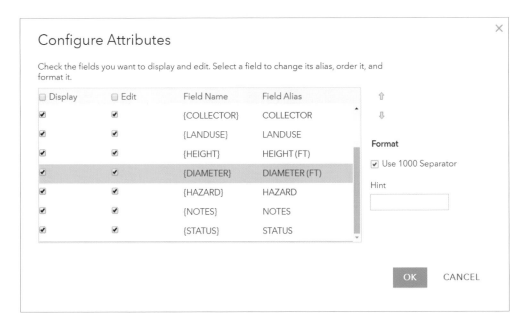

Configure Attributes   ✕

Check the fields you want to display and edit. Select a field to change its alias, order it, and
format it.

☐ Display	☐ Edit	Field Name	Field Alias	
☑	☑	{COLLECTOR}	COLLECTOR	⇧
☑	☑	{LANDUSE}	LANDUSE	⇩
☑	☑	{HEIGHT}	HEIGHT (FT)	
☑	☑	{DIAMETER}	DIAMETER (FT)	
☑	☑	{HAZARD}	HAZARD	
☑	☑	{NOTES}	NOTES	
☑	☑	{STATUS}	STATUS	

Format

☑ Use 1000 Separator

Hint

OK   CANCEL

6. **Click the STATUS row, and then click the Move up arrow several times until it is above LANDUSE. Then click OK.**

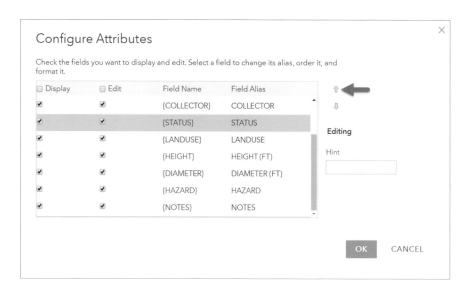

7. **At the bottom of the Configure Pop-up pane, click OK.**

8. **Click a tree feature to view its pop-up window.**

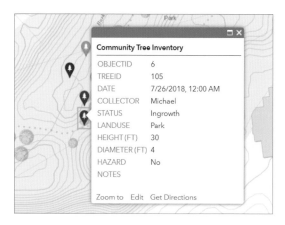

9. **Save the map.**

10. **Leave the map open.**

TIP *You can configure pop-up windows at any time. Depending on your needs, you can customize pop-up windows to include photos and charts.*

Share the map and enable editing

The map is ready to be used in the field. The way this map is shared and the permission is set determines who in your organization can view it only or collect and edit the information. A collector who has editing permission can adjust the data collection form in the field. Otherwise, you can give permission to collectors to collect only. First, you will adjust some map properties.

1. **In the same map, under Details, click the About this Map button, and then click More Details.**

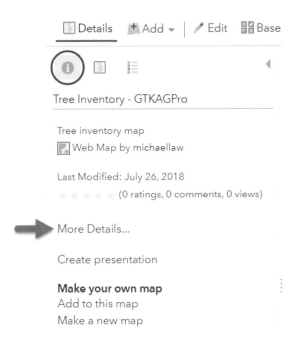

2. **On the Settings tab, scroll through the General settings section until you see Delete Protection. Click the "Prevent this item from being accidentally deleted" check box to enable it. Click Save.**

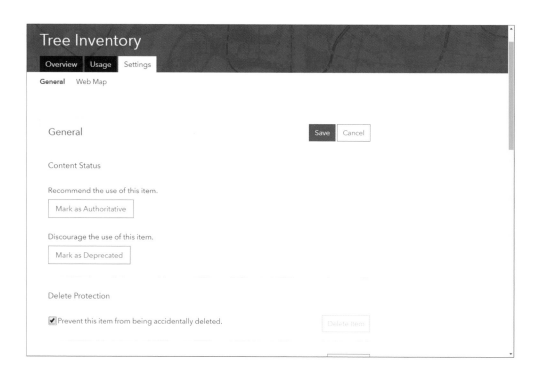

3. **Scroll down to the Web Map settings section until you see Save As. Click the Save As check box to clear it. Click Save.**

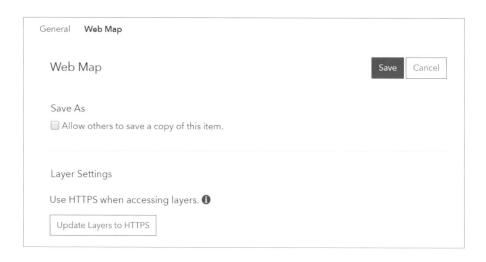

Enabling delete protection will prevent you or another member of your organization from mistakenly deleting the map. Disabling the Save As setting will prevent members of your organization from saving a copy of the map, effectively preventing multiple copies.

4. **Scroll down to the Application Settings section. Under the Find Locations setting, select the By Layer check box, and click the Add Layer button.**

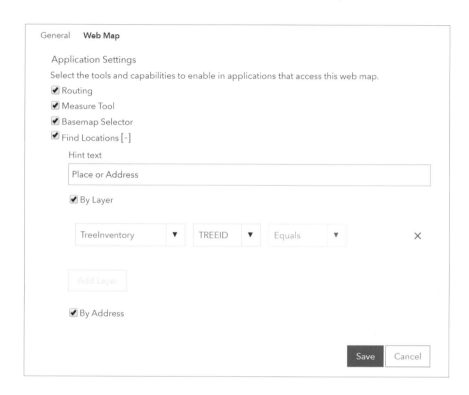

You can search the Trees layer based on the COLLECTOR attribute, allowing you to search which data collector in the field has submitted features.

5. **Click Save.**

6. **On the Overview tab, click Share.** The Share dialog box provides options to share to your organization or to groups to which you may belong. Do not worry if your Share dialog box looks different from the graphic.

TIP To continue the exercise, you do not have to share the map with anybody.

TIP If you belong to a group, you can share the map by selecting the check box for the group. You may be prompted to update layer sharing to match how the web map is shared.

7. **Click OK.** Next, you will enable editing for the TreeInventory layer so that it can be accessed in the Collector app.

8. **On the main navigation bar, click Content.**

9. **View item details for the Tree Inventory feature layer. (Make sure you choose the correct item—the feature layer, not the web map or service definition.)**

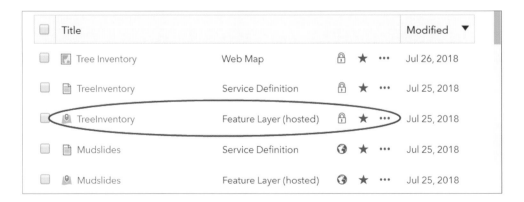

10. On the feature layer detail page, click the Settings tab.

11. Scroll to the Feature Layer (hosted) section.

12. Click the "Enable editing" check box to enable it.

13. Ensure that the "Add, update, and delete features" option is selected in the "What kind of editing is allowed?" section.

14. Finally, click the "Keep track of who created and last updated features" check box.

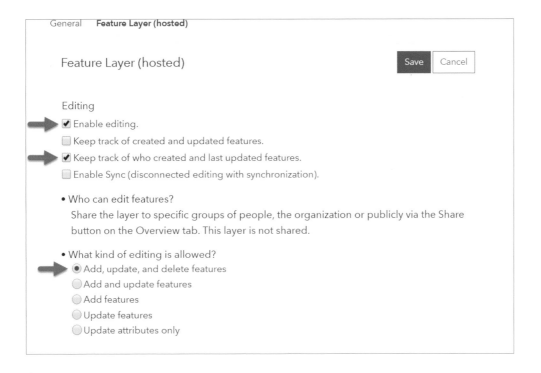

Enabling editing allows you and your organization to edit features on the map through the Collector app. Enabling tracking will help you keep track of who is making changes to features.

15. **Click Save.**

16. **Open the Tree Inventory web map, and click Save.** When a map is created with a feature layer that is uneditable, you must open the map and save it again after the feature layer is changed to allow editing.

Your map is now ready for data collection in the field.

EXERCISE 6C

Collect data using Collector for ArcGIS

Estimated time to complete: 25 minutes

TIP *This exercise requires you to use the Collector for ArcGIS app on your mobile device. To install the app, visit the appropriate store for your device platform:*

- *For IOS devices, go to the App Store: itunes.apple.com/us/app/arcgis/id589674237*
- *For Android devices, go to Google Play: play.google.com/store/apps/details?id=com.esri.arcgis.collector*

Exercise workflow

- *Open the Collector for ArcGIS app on your mobile device.*
- *Collect tree locations, and enter tree inventory data.*
- *Open ArcGIS Pro to view the map and add additional points.*
- *Use the Collector app to collect tree data in the field.*

Open Collector for ArcGIS

1. **Open Collector for ArcGIS on your mobile device.**

2. **Tap ArcGIS Online to sign in to your organization.**

3. **Sign in with your ArcGIS Online user name and password.**

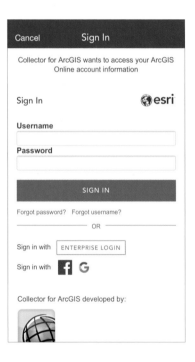

4. **On the All Maps or My Maps screen, tap the listed Tree Inventory map to open it.**

5. **If your location is not automatically detected (it should be), tap the My Location button to locate yourself on the map.** Location services must be enabled on your mobile device.

   Your location is acquired by the device's GPS and is displayed by the dot at the center of the map. If you pan away from your current location, tap My Location at any time to return to it. Do not worry if your map location does not match the ones in the exercise graphics—your location will differ from the ones shown in the graphics.

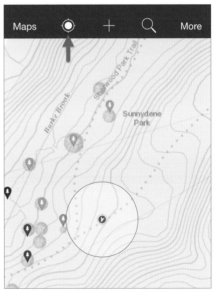

Collect a tree location and enter the data

You are ready to collect a feature location and create tree inventory data within the Collector app. For this exercise, assume that you are at the location of a tree to collect its location and tree inventory data.

1. **Tap Collect New.**

> **TIP** *If you are using an iPad or Android device, some icons and workflows will differ slightly.*

2. **Tap Planted.** A tree inventory is created for a planted tree. The form is the same as the one you used in the ArcGIS Online map. The Location heading displays the latitude and longitude, as well as an indicator that shows how accurate the GPS is (in this case, 65 meters, although your indicator may be different).

3. **Tap Collect Settings.**

The settings show the required location accuracy and interval required for collection. The required accuracy is currently set to 100 meters, and the streaming interval is set to 5 seconds. Changing these settings to lower values increases the collection accuracy.

4. **Tap Done to close the settings screen.**

5. **Tap Map.**

6. **Tap a different location on the map to choose it.** An orange dot shows where you tapped. You can use this feature to set a new location for your tree, or if the initial GPS location is not accurate enough, to fine-tune the location. Don't worry if your screen looks different from the images shown. For the purposes of this exercise, you can collect data for any location.

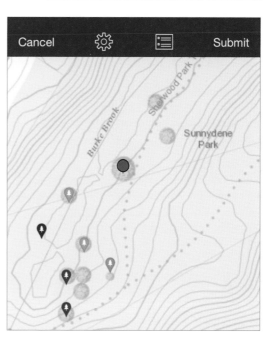

TIP *If you want to use your exact location again, tap the My Location button in the lower-left corner of the screen. You can also tap and hold to magnify the area around the location and release to select it.*

7. **Tap Collect Attributes.**

8. **On the tree inventory form, tap the TREEID field, and type** 201.

9. **For the other fields, enter the following details:**
    - **DATE: (today's date)**
    - **COLLECTOR:** (your name)
    - **STATUS: Planted**
    - **LANDUSE: Park**
    - **HEIGHT:** 20
    - **DIAMETER:** 3
    - **HAZARD: No**

    When you are finished completing the attribute fields, you are ready to save the location.

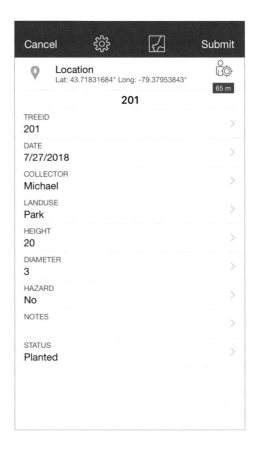

10. **Tap Submit.** It takes a few seconds to submit your tree inventory location. The data is saved to your ArcGIS Online organization. If you shared your map with others in your organization, it is available to everybody who has access to it.

ON YOUR OWN

Feel free to create more tree inventory locations, either by specifying locations manually or physically moving to another area and using the GPS to find your location.

11. **Tap a tree inventory location you just created. (If the location is difficult to see, tap My Location to disable the GPS.)** Results appear for the location you tapped. If there are multiple tree inventory results, they are listed.

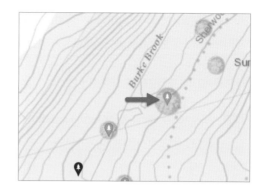

12. **Tap a result.** The attribute details are listed.

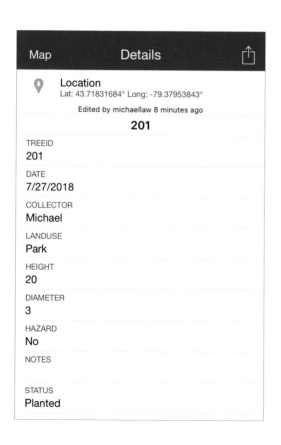

Map   Details

Location
Lat: 43.71831684° Long: -79.37953843°

Edited by michaellaw 8 minutes ago

201

TREEID
201

DATE
7/27/2018

COLLECTOR
Michael

LANDUSE
Park

HEIGHT
20

DIAMETER
3

HAZARD
No

NOTES

STATUS
Planted

The information is not editable on the Details screen. However, you can tap the button to the right of the Details title to copy, edit, delete, zoom to, or find directions to the location.

13. **Tap Map to return to the map.**

> **TIP** *Tap More to access bookmarks, hide or show layers, measure distance or area, and change basemaps.*

View the map in ArcGIS Pro

After collecting data in the field, you can return to ArcGIS Pro to see all of the tree inventory data.

1. **In ArcGIS Pro, open TreeInventory.aprx, found in your \EsriPress\GTKAGPro\ TreeInventory folder.**

2. **In the Catalog pane, click Portal, right-click the Tree Inventory map, and click Add and Open.** The Portal tab in the Catalog pane lists all of the maps and layers that are stored in your ArcGIS Online organizational account. (The contents in your account may look different, depending on the contents saved to your account.)

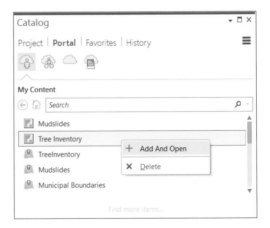

3. Click a tree inventory location that you created using the Collector app.

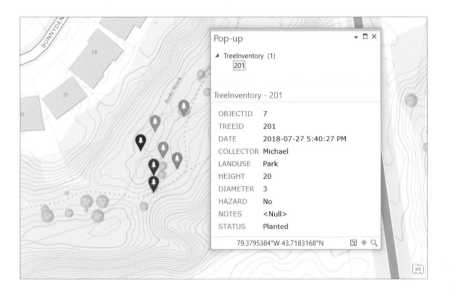

A pop-up window shows the attribute information for the location.

4. Open the Trees attribute table.

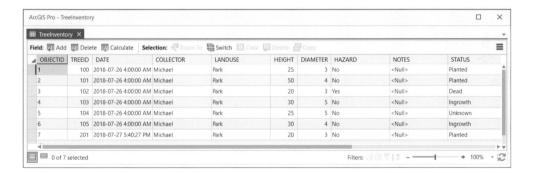

The attribute table lists all of the features and data you collected. The attribute data is editable, and when changes are saved, they are reflected in ArcGIS Online and the Collector app.

Next, you will add a location using edit tools.

5. On the Edit tab, click Create.

6. **In the Create Features pane, click Planted.**

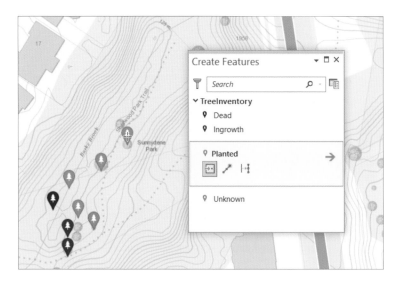

By default, the Point tool is chosen and allows you to create a point feature anywhere on the map. The cursor changes when it is over a map to a cross hair with the Planted symbol.

7. **Click a location on the map to place a Planted point.**

A Planted tree point is added to the map. You can quickly create points for a large area using this method. For each tree point added to the map, a corresponding record is also added to the attribute table.

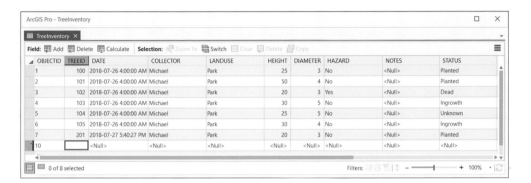

8. **Close the Create Features pane.**

9. **In the Trees attribute table, click the last row. Click the TREEID field, and type** 202. **Click the DATE field, and select today's date.**

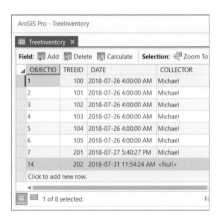

Green coloring of the row and column headings indicates which cell is currently being edited. The row is highlighted in aqua, indicating that it is selected.

10. **Enter details for the remainder of the attributes in the row. When you are finished entering the last attribute field, press Enter.**

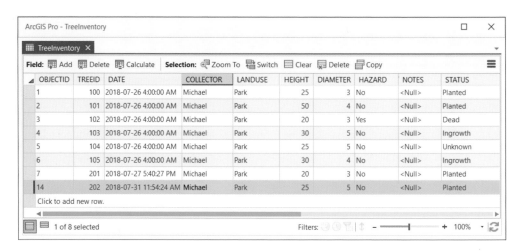

TIP *You can also edit any of the existing tree inventory points.*

The attribute data is saved.

11. **Close the attribute table.**

12. **On the Map tab, click Explore. Click the newly created tree inventory location to see its pop-up window.**

Because the map was opened from the Portal tab in the Catalog pane, the data is live and linked to your ArcGIS Online organization. Thus, any changes made can be immediately seen in both the ArcGIS Online map view and the Collector app.

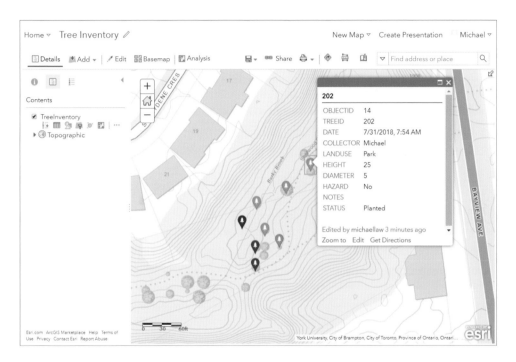

Tree Inventory map in ArcGIS Online.

Tree Inventory map in Collector for ArcGIS.

13. **Save the project.** You are finished with this chapter. If you are not continuing to chapter 7, you can close ArcGIS Pro.

Summary

You now have some experience in using the ArcGIS platform—using ArcGIS Pro on the desktop, ArcGIS Online on the web, and Collector for ArcGIS on a mobile device. By creating the tree inventory collection system, you have created a management tool for managing individually collected tree data. This tree inventory can grow to provide information about a population of trees, typically known as an urban forest. It can be used to provide more information about the species of trees that cover an area and to calculate canopy cover. A tree inventory map enables urban forest managers to identify areas in which tree conditions can be poor and so prioritize maintenance.

By now, you should feel comfortable with preparing and creating data in ArcGIS as well as using the ArcGIS Online platform. Remember, an ArcGIS Online organizational account is required to share maps between ArcGIS Pro, ArcGIS Online, and the Collector app; all edits made to web layers are live and immediately stored. You can organize and manage your content in your ArcGIS Online organization at any time.

Glossary terms

organization
data type
alias
web layer

## CHAPTER 7

Geoenabling your project

Exercise objectives

7a: Prepare project data

- *Join a table*
- *Create and calculate a field*
- *Symbolize using graduated colors*

7b: Geocode location data

- *Create an address locator*
- *Geocode addresses*
- *Rematch addresses*

7c: Use geoprocessing tools to analyze vector data

- *Create buffers*
- *Clip features*
- *Create a spatial join*
- *Calculate distances*
- *Select by location*
- *Dissolve features*
- *Intersect layers*

In chapter 4, you learned how to create and modify features. You can also create features from information that describes or names a location—typically an address—through a process called *geocoding*. When working on projects, you are often faced with situations in which data is not readily available. One of the advantages of geocoding is the ability to quickly create data from a list of points of interest that you have collected or researched.

You need several components to geocode addresses:

- An address table—a list of addresses for features that you want to map and that are stored as a database table or a text file

- Reference data—commonly from a streets layer, on which the addresses can be located
- An *address locator*—a file that contains the reference data and various geocoding rules and settings

The ArcGIS Pro geocoding service uses address information in the attribute table of the reference data to calculate where addresses from the address table are located and creates points at those locations. As with other GIS data, the higher the quality of the reference data, the more accurately addresses can be geocoded. Ideally, street-level reference data includes attributes such as street name, street type, postal or ZIP Code, and directional prefixes and suffixes to avoid ambiguity in address locations. And each street is divided into line segments that have starting and ending address numbers so that you can determine separate address ranges for each side of the street.

Address locators come in many styles; your reference data will determine which style can geocode your address table appropriately. In ArcGIS Pro, address locators are created in the Geoprocessing pane. For more information on geocoding, go to ArcGIS Pro Help > Data > Data types > Geocoding > Get started > What is geocoding?

Scenario: affordable housing for low-income residents in many cities is often isolated from basic and essential social services. Using GIS to visualize the current state of community or public housing sites can help determine where potential services, such as public transportation, child care, food assistance, libraries, medical clinics, and job-training facilities, can be planned and built with improved access. In this chapter, you are an analyst who will prepare affordable-housing data for use in support of policy planning and consultation. You will use several important tools to prepare the data, which will include joining tables from other sources and spatially joining data. You will also geocode data and use *overlay* analysis tools to identify and visualize social service sites that are near public housing sites.

GIS IN THE WORLD: SOCIAL SERVICES

Mapping urban areas can help cities target policies that are most efficient and effective for their communities, particularly for those who are less fortunate. However, finding a solution to a problem such as homelessness entails understanding the associated issues. GIS has become fundamental to that process.

Homelessness prevention should, of course, be the first priority. But when homelessness persists, it becomes necessary to have a structure ready to supply fundamental care and services. Finding a location for shelters gets to be a difficult situation for local governments, because businesses may find it undesirable to have homeless people close by and, therefore, may resist their accommodation. Read more about how social scientists at Morgan State University studied the impact of establishing a permanent homeless facility in a Baltimore, Maryland, neighborhood, "Mapping Urban Inequalities with GIS": (www.esri.com/news/arcnews/spring10articles/mapping-urban.html).

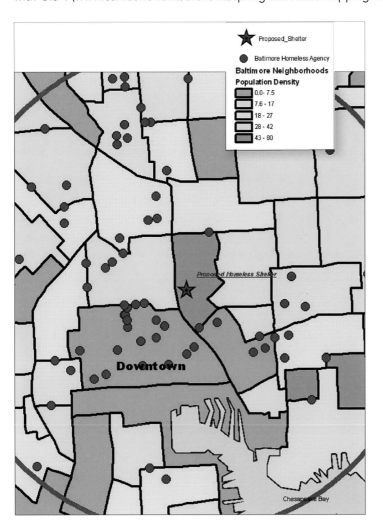

This map of neighborhoods in Baltimore shows the population density for defined neighborhood boundaries and locations of current service providers within a 1.5-mile radius. Within this radius are at least 60 percent of the providers of services to the homeless. Courtesy of Linda Loubert, Morgan State University, Baltimore, MD.

DATA

In \\EsriPress\GTKAGPro\CommunityHousing\CommunityHousing.gdb:

- Social services—a point feature class of existing social services (*Source: Los Angeles County Enterprise GIS*)
- LA census tracts—a polygon feature class of census tracts in Los Angeles County (*Source: Los Angeles County Enterprise GIS*)
- City boundaries and annexations—a polygon feature class of the city boundary of Los Angeles (*Source: Los Angeles County Enterprise GIS*)
- Streets—a line feature class of streets (*Source: Los Angeles County Enterprise GIS*)

In \\EsriPress\GTKAGPro\CommunityHousing:

- Median household income (Los Angeles County, 2012 American Community Survey five-year estimates)—a text file that contains median household income values for the County of Los Angeles (*Source: American FactFinder*)
- Public housing—a text file that contains a compiled list of community or public housing units (*Sources: Los Angeles County Enterprise GIS; Housing Authority of the City of Los Angeles; Housing Authority of the County of Los Angeles*)

EXERCISE 7A

Prepare project data

Estimated time to complete: 20 minutes

In this exercise, you will prepare census data for use later in the analysis.

Exercise workflow

- *Add the median household income table to the project.*
- *Review table information.*
- *Join the median household income table to the census tracts layer.*
- *Create a new field, and calculate average median income.*
- *Symbolize census tracts using graduated colors.*

Join a table

US Census data is a reliable and helpful resource for any social research project and is collected with geographic detail that ranges from fine to general granularity. Almost all of the data is stored as values in tables and not geoenabled. However, you can perform several steps to turn this data into a map. First, you will view household income data provided by the US Census Bureau.

1. **In ArcGIS Pro, open CommunityHousing.aprx from the \\EsriPress\GTKAGPro\ CommunityHousing folder.** The project opens to a map of Los Angeles County census tracts and a basemap.

2. **Add the household_income.csv file to the project from \\EsriPress\GTKAGPro\ CommunityHousing.**

3. **Open the household_income table to review it.** The table has several columns, including Geo.id and Geo.id2. These names are standard identifiers that the US Census Bureau provides to identify census tracts. The HHINCOME column represents household income (in US dollars). However, knowing only the household income is not so useful on its own. You will join it to the Los Angeles County census tracts layer to make the information spatial before you calculate the area median income.

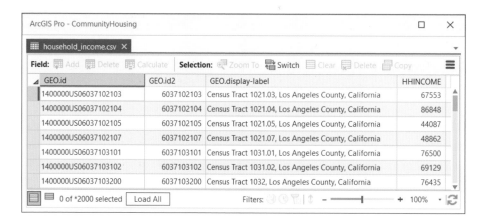

4. **Open the la_census_tracts attribute able to review it.**

5. **Join la_census_tracts to household_income.csv. Base the join on the GEOID2 and Geo.id2 fields, respectively. You can ignore the index warning.** (Remember, you learned how to join tables in chapter 4.)

REMIND ME HOW

1. *Go to the la_census_tracts attribute table pane options menu, point to Joins and Relates, and click Add Join.*

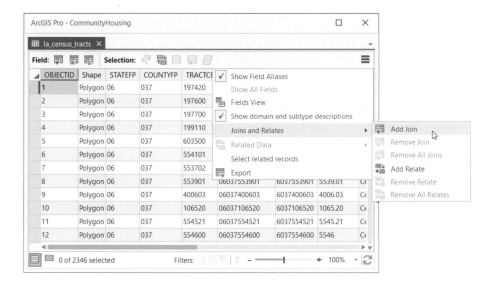

2. *In the Geoprocessing pane, complete the details for the Add Join tool. For Input Join Field, select GEOID2. Make sure the Join Table is set to household_income.csv. The Output Join Field should automatically populate as Geo.id2.*

3. *Run the tool.*

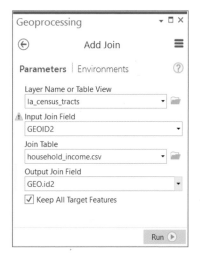

6. **In the attribute table, scroll to the right to view the joined attributes from the household_income.csv file.** Now you can see the household income for each census tract in Los Angeles County in the HHINCOME column. Keep the la_census_tracts attribute table open.

Create and calculate a field

Local government uses a county's area median income (AMI) to determine a household's eligibility and selection for public housing. Each year, the US Department of Housing and Urban Development (HUD) estimates the median household income of an area so that household incomes can be expressed as a percentage of the AMI and divided into income categories.

1. **In the la_census_tracts attribute table, open the Fields View from the attribute table pane options menu.** The Fields table opens on a new tab.

2. **At the bottom of the Fields table, click the row titled Click here to add a new field.**

3. **When the Field Name box is in focus, enter AMI, and then press the Tab key to move to the next box.**

4. **In the Alias box, enter** Area Median Income.

5.   Change Data Type to Double.

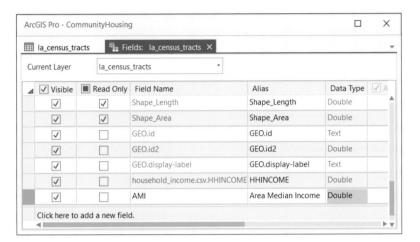

6.   Change Number Format to Numeric. Under Rounding, change the number of decimal places to 2, and click OK.

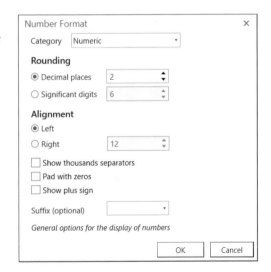

7.   On the Fields tab (on the ribbon), click the Save button.

8.   In the la_census_tracts attribute table, scroll to the right, and right-click the Area Median Income column heading. Then click Calculate Field. AMI is calculated by dividing the household income by median income. Based on US Census Bureau statistics, the median income of Los Angeles County is $56,241 (US Census Bureau: 2008 to 2012).

9. **In the Geoprocessing pane, under Expression, scroll through the Fields list, and double-click HHINCOME. Click the / button, and enter** 56241 * 100 **in the expression box.** The equation la_census_tracts.AMI = !household_income.csv.HHINCOME! / 56241 * 100 is ready to be calculated. The value is multiplied so that it can be represented as a percentage.

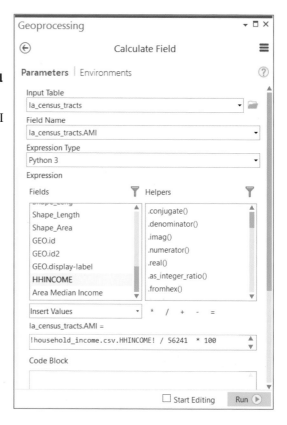

10. **Click Run.**

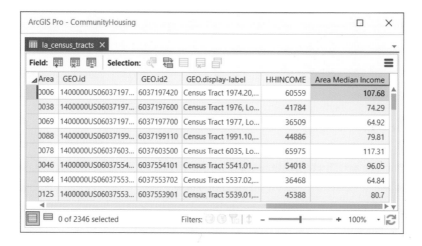

Applying a meaningful symbology scheme to the AMI data will allow you to focus on areas of low or very low income. You will add the symbology next.

Symbolize using graduated colors

With area median income percentages calculated for each census tract, you can now see which areas have income limits that determine eligibility for a variety of HUD housing programs. For example, an eligible household's income may equal 80 percent of the area median income, a common maximum income level for participation in federal and state programs. A common classification used for AMI percentages is as follows:

- Extremely Low Income: less than or equal to 30 percent of AMI
- Very Low Income: 31 percent to 50 percent of AMI
- Low Income: 51 percent to 80 percent of AMI
- Moderate Income: 81 percent to 120 percent of AMI
- High Income: greater than 120 percent of AMI

You will symbolize the data based on these categories.

1. **Close any open tables. With the la_census_ tracts layer selected, go to the Appearance tab, open the Symbology drop-down list, and select Graduated Colors.** A random color is applied to the census tracts. You will change the symbology settings to categorize the area median income percentages.

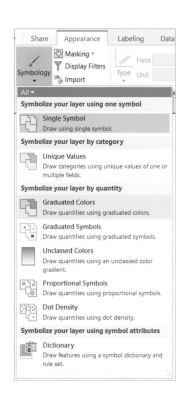

2. **In the Symbology pane, from the Field drop-down list, select Area Median Income, and from the Method drop-down list, select Manual Interval.** The Manual Interval method allows you to create class breaks based on custom values.

3. **Select Orange-Red (5 Classes) from the Color scheme drop-down list.**

4. **In the Classes tab, click the More down arrow, and click Reverse symbol order. (Make sure that you do not reverse the order of the actual class breaks.) Alternatively, right-click Symbol, and click Reverse symbol order.**

5. **In the class breaks table, click the first field in the first Upper value column, type** 30, **and press Enter. In the corresponding Label field, enter** Extremely Low Income.

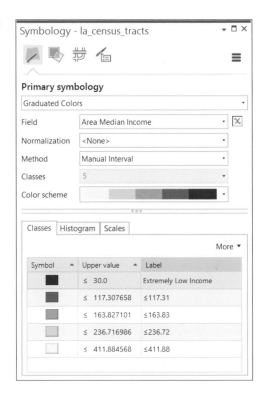

6. **Continue to enter the upper values and labels for the remaining categories:**
   - **Upper value: 50; Label:** Very Low Income
   - **Upper value: 80; Label:** Low Income
   - **Upper value: 120; Label:** Moderate Income
   - **Upper value: 412; Label:** High Income

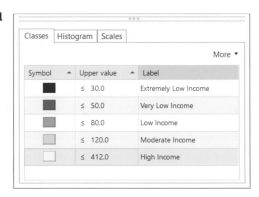

7. **Click the More down arrow, point to Symbols, and click Format all symbols. Click Properties, and change the outline color to Soapstone Dust. Click Apply.**

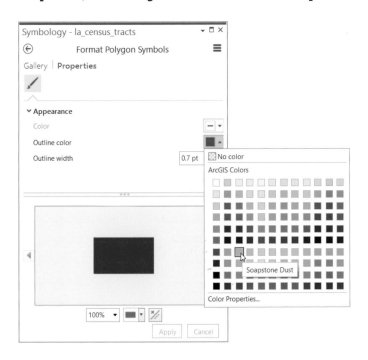

Reversing the order of the symbols reverses the color scheme so that the areas of interest—extremely low income, very low income, and low income—are symbolized with a darker shade of red and emphasized. You can already see clusters of census tracts that are classified as lower income on the basis of the area median income. Some census tracts may not have any symbology, which is often the case for areas designated as parks, airports, or port lands.

8. **Save the project.**

9. **Keep the project open, and continue to exercise 7b.** This layer has been prepared for further analysis. Before you perform more geoprocessing, you will add more layers by geocoding data.

ON YOUR OWN

Adjust the symbology to your liking. Because areas that have moderate and high income as a percentage of area median income will not be further included in the analysis, you can symbolize these two classes with no fill and no outline. Also, to learn more about data classification in ArcGIS Pro, go to ArcGIS Pro Help > Maps and scenes > Layers > Layer properties > Symbology > Data classification methods.

EXERCISE 7B
Geocode location data

Estimated time to complete: 30 minutes

Although an address allows someone to easily find a location of a business or home while walking or driving, a GIS requires similar information to achieve the same result. A person must know two main pieces of information: the address of the business or home and knowledge or reference to where that address is, such as street, main intersection, or neighborhood. Likewise, a GIS needs map or street information to know where to create address points. The process of creating map features from addresses, place-names, and similar information is called *geocoding*.

For your community housing project, a list of public housing sites—you must add a simple text file with addresses, names, and other information to your map. You will use a streets layer of Los Angeles County as the reference data—the spatial data that contains information about all of the addresses. An address contains certain address elements (such as address number, street type, direction, and ZIP Code) and is presented in a range of formats. When you geocode the table of public housing sites, you will use the reference data to create an address locator to create point features that represent the locations of the addresses (also known as *locator styles*). When geocoding your table of addresses, the address is interpreted and compared to the address locator.

Exercise workflow

- *Create an address locator based on Los Angeles streets.*
- *Geocode addresses of public housing sites.*
- *Interactively rematch public housing addresses that cannot be geocoded.*

GEOCODE USING THE ARCGIS® WORLD GEOCODING SERVICE

Often, obtaining accurate reference data can be difficult. The ArcGIS World Geocoding Service references rich and high-quality point and street address data, places, and gazetteers that cover much of the world. The service is available for users who are signed into ArcGIS Online and operates under a credit-based usage model.

Create an address locator

An address locator defines which reference data will be used to search for addresses, which elements of the reference data will be searched against, and what format the address data will use. ArcGIS Pro comes with several predefined address locator styles that you can use to create address locators. An address style includes the elements of the address that are present (such as street name, number, direction) and the elements that will be used in locating your addresses.

Each address locator style is unique and has specific requirements regarding the reference data that it can use to match the format of addresses being geocoded. Some address locators have more requirements than others, which allows for a wide range of accepted address formats. Using the appropriate address locator style depends on knowing what type of reference data you have and what format your addresses will use. In this case, you will use a US Dual Range address style.

1. **Continue working with the map from exercise 7a. On the Analysis tab,** **click the Tools button.**

2. **In the Geocoding Tools toolbox, search or navigate for the Create Address Locator tool. Open the tool.**

3. **In the Geoprocessing pane, select US Address – Dual Ranges from the Address Locator Style drop-down list. Click the Browse button next to the Reference Data box, and select the streets layer from the Community Housing geodatabase.**

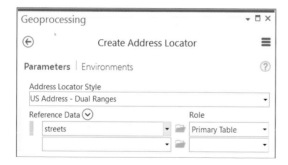

The streets layer acts as the reference data. Because you are using only one layer as reference data, it is assigned the role of Primary Table.

The Field map automatically populates with fields from the streets layer that match the fields that the address locator style requires. Although most of the required fields, indicated by an asterisk (*), are mapped, some fields may not be.

4. **Scroll down the field map to complete these steps.**
   - **For the Prefix Direction field, from the Alias Name drop-down list, select PREDIRABRV.**
   - **For the Prefix Type field, for Alias Name, select PRETYPABRV.**
   - **For Suffix Type and Suffix Direction, select SUFTYPABRV and SUFDIRABRV, respectively.**

5. **Click the Browse button next to the Output Address Locator box. Save the address locator as** Housing_Address_Locator **to your project folder. (Make sure you are not trying to save it to the geodatabase; address locators must reside in folders.)**

6. **Click Run to create the address locator.** Creating the locator will take a few moments as ArcGIS Pro analyzes the streets layer to create the address locator. You can view the progress at the bottom of the Geoprocessing pane. When the address locator is complete, it is automatically added to the project. In the Catalog pane, expand the Locators folder to verify that the locator was successfully added. The other address locator—ArcGIS® World Geocoder—is included with ArcGIS Pro and is configured to help you locate places on a map in situations in which you may not be able to create your own.

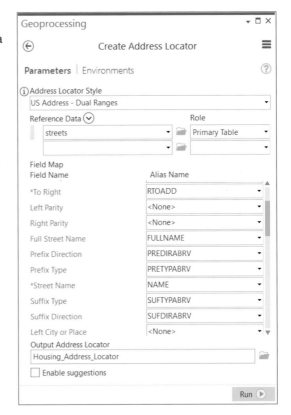

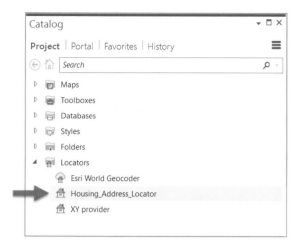

An address locator also determines other settings for geocoding, including what comprises a matched address, search parameters, and tolerance for spelling errors. You will see more of these settings next.

USING THE LOCATE PANE

You can use the Locate pane (on the Map tab) to find addresses using the ArcGIS World Geocoding Service without the need for creating an address locator. This pane is useful for finding addresses one at a time or finding places for which you may not know the address, such as a business or public place. The Locate pane does not find addresses in any custom address locators that you create though. To learn more about the Locate pane, go to ArcGIS Pro Help > Data > Data types > Geocoding > Locate addresses > About finding an address.

Geocode addresses

Now that the address locator is ready, the next step is to geocode the list of public housing sites, which is provided in your project folder. The output of the geocoding process can be either a shapefile or a geodatabase feature class of points. The geocoded data has all the attributes of the address table; some of the attributes of the reference data; and optionally, some new attributes, such as the x,y coordinates of each point.

1. **In the Catalog pane, expand the CommunityHousing folder, and add the public_housing_sites.csv file to the current map. Open the table.** The table contains many attribute columns, including the addresses of 91 public housing sites, the agencies that manage or own them, a brief description, and the number of units at each site. The address information is added during the geocoding process.

2. **In the Contents pane, right-click public_housing_sites.csv, and click Geocode Table. In the Geocode Table pane, read the guided workflow, scroll down, and click Go to Tool.**

3. **For Input Table, make sure public_housing_sites.csv is already selected.**

4. **Click the Browse button next to the Input Address Locator box. Browse to Project > Folders > CommunityHousing, and select Housing_Address_Locator.**

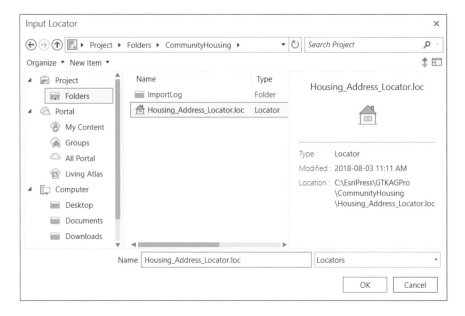

5. **Click the Browse button next to the Output Feature Class box. Browse to Project > Databases > CommunityHousing, and enter the name as** Public_Housing_Sites. **Click OK.**

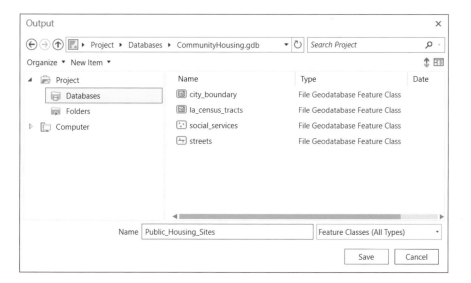

The Geocode Table settings are set and ready to be run. Your display should look as follows:

6. **Run the tool to geocode the public housing sites.**

Geocoding Completed   ×

87 Matched (95.60%)
3 Unmatched (3.30%)
1 Tied (1.10%)

Average speed: 192366 (records/hour)

Start rematch process?

Yes ·   No

A dialog box appears that shows what addresses from the public housing sites table are matched and unmatched with the streets data. All but three addresses have been matched. You'll perform an address rematch in the next exercise.

7.   Click No to close the geocoding results dialog box.

8.   Zoom the map so that you can see the public housing sites better.

The public housing sites feature class is added to the map when the geocoding is complete. You can begin to see that there is a visual correlation between the location of public housing sites and areas that are classified as extremely low income.

9. **Open the Public_Housing_Sites layer attribute table.** The table contains 91 records, one for each public housing site that is in the public_housing_sites.csv file. The Status attribute column tells you if each address is matched (M), unmatched (U), or tied (T). The Score attribute column tells you how well it scores when comparing the table to the reference data. The Match_type attribute shows how an address was matched—in this case, automatically (A) matched. The Match_addr attribute shows the actual address to which the public housing site was geocoded, and the Side attribute tells you whether an address is on the right or left side of the street. The original fields of the table are preserved and include the prefix USER_.

ON YOUR OWN

Sort the Public Housing Sites table to see which addresses are matched, unmatched, or tied. Review the addresses to guess why they may not have been geocoded. If you want, use the Locate pane to see if the ArcGIS World Geocoding Service can find the addresses you are reviewing. (Do not actually add points though. You will interactively rematch unmatched addresses next.)

Rematch addresses

Several public housing sites were not geocoded. Next, you will use the Rematch Address feature to identify the unmatched addresses and interactively rematch them.

1. **Turn off the la_census_tracts layer (the Topographic basemap lets you see street names).**

2. **In the Contents pane, right-click the Public_Housing_Sites layer (not the table), and click Data > Rematch Addresses to open the Rematch Addresses pane.** The Rematch Addresses pane allows you to review which addresses are unmatched or insufficiently matched, and then go through the parameters for each address and adjust them to make a match. When rematching addresses in this exercise, your view may change to look like the next image. To return to the original view of the points, in the Contents pane, right-click Public_Housing_Sites, and click Zoom To Layer.

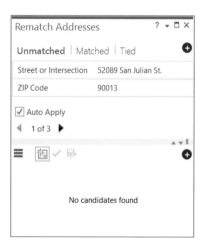

The Rematch Addresses pane displays the first of three addresses that are unmatched. Possible candidates are listed in the lower portion of the pane. However, the first address has no candidates. Make sure that Auto Apply is checked so that any changes in the pane are applied without the need to click the Apply button (which will speed up the candidate search). You can use the arrows to browse through each address individually in any of the three categories: Unmatched, Matched, or Tied.

3. **In the Rematch Addresses pane, for the first address, change the Street or Intersection address to** 520 San Julian St., **and press Enter.**

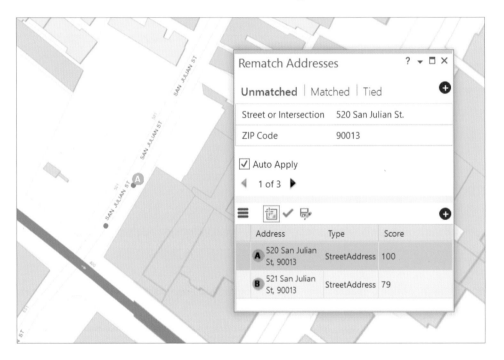

Two candidates are found and are listed in the lower portion of the Rematch Addresses pane: candidate A, 520 San Julian St., with a score of 100; and candidate B, 521 San Julian St., with a score of 79. Clicking the candidate shows its location on the map.

TIP *If your Topographic base layer is not showing when zoomed in close, you can adjust its visibility. In the Contents pane, double-click the Topographic layer. Change the In beyond (maximum scale) value to **1:125**, and click OK.*

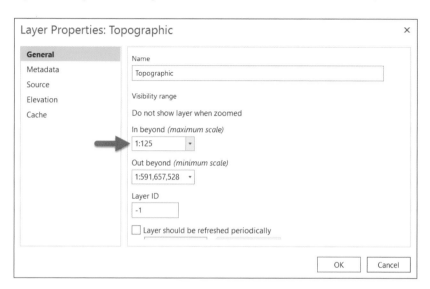

4.   **Click the menu button, and then click Map All Suggestions.**

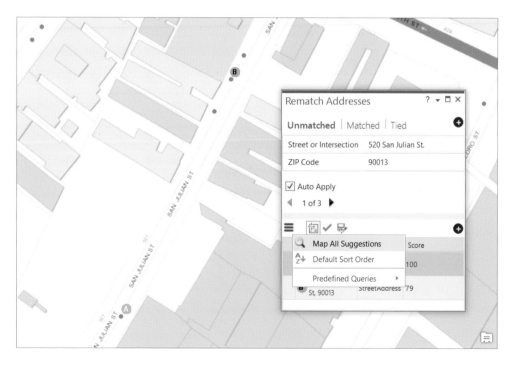

This option adjusts the zoom of the map so that both candidates are visible within the map extent. This visibility makes it easier to check the addresses of other public housing sites nearby.

5. **On the Map tab, click the Explore down arrow, and select Topmost Layer. Click the point nearest to candidate A.**

The pop-up shows the address as 518 San Julian St. for the selected point. This matched address can be considered verification that candidate A, 520 San Julian St., is indeed located nearby and on the same side of the street. You can click the points around candidate B to see that the addresses are uneven numbers and considered not as close to candidate A.

6. **Make sure that candidate A is selected in the Rematch Addresses pane, and click the Match button.** The Rematch Address pane is updated, and now there are only two unmatched addresses. The ZIP Code box shows and is the reason why the geocode cannot find a match, because of missing information. You may have previously used the Locate pane to find addresses; if you have used it before, the next step should look familiar.

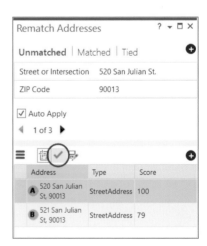

7. **On the Map tab, click Locate.**

8. **In the Locate pane, type** 417 E 7th St, Los Angeles, California, USA. **As you enter the address, the box auto-completes and displays suggested addresses that may match.**

   The 90014 ZIP Code matches the address that you are looking up.

9. **Enter** 90014 **into the ZIP Code box of the Rematch Addresses pane.**

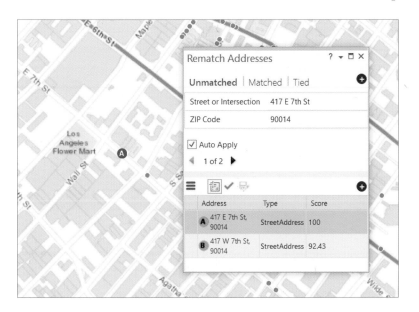

Notice the difference in the results and the colors of the candidate icons: green for the Locate pane results and orange for the Rematch Addresses pane. Two candidates appear, and candidate A has a match score of 100. Because the original address has the east (E) component of the address, candidate A is the best match. You may also notice that the result from the

Locate pane may not be in the same location as the candidate from the Rematch Addresses pane. The reason is because the reference data is different—the streets layer that you used as reference data to create the address locator is slightly different from the reference data used by ArcGIS Online to rematch and geocode addresses.

10. **Click the Match button.** Finally, match the third unmatched address. The listed address—4909 Merrianwood Dr.—is incorrect. The correct address provided for you is actually 4909 Marionwood Dr.

11. **In the Rematch Addresses pane, type** 4909 Marionwood Dr. **in the Street or Intersection box.**

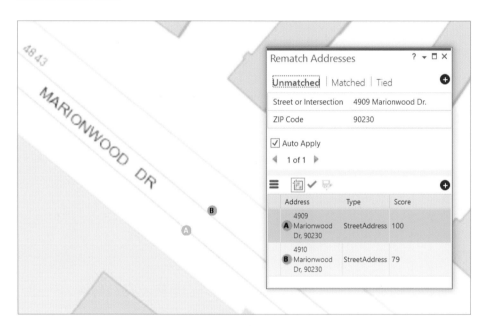

12. **Click the Match button.** All of the unmatched results are now matched. The Unmatched category in the Rematch Addresses pane disappears, and only the Matched and Tied categories remain. The existence of a Tied category means that there are candidates with the same match score.

13. Click the Tied category.

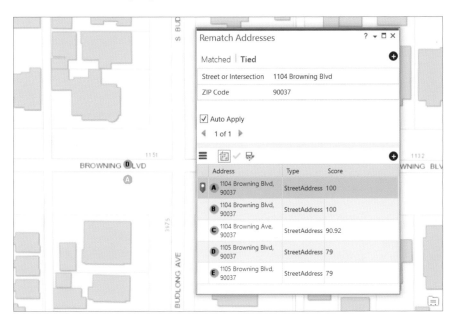

One address in the record lists two candidates with the same match score of 100. The green placemark icon to the left of the candidate A icon indicates that it is currently mapped. Both candidates have the same address. However, they are in different locations on the map.

14. In the Locate pane, look up the 1104 Browning Blvd **address. Click the green A results to select them on the map.**

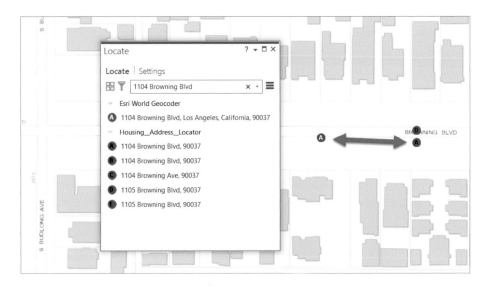

15. **Click the Rematch Addresses pane. Zoom out slightly to see the results, or from the pane options menu, use Map All Suggestions.**

It seems as if 1104 Browning Blvd result A in the Locate pane more closely matches candidate B in the Rematch Addresses pane. Remember that although the reference data is different, there is a close spatial relationship between the two.

16. **In the Rematch Addresses pane, select candidate B, and click the Match button.** The Tied category disappears, and all 91 public housing sites are now considered matched.

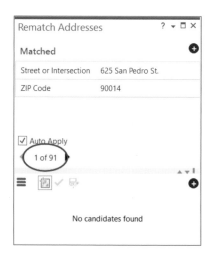

17. **On the Edit tab, in the Manage Edits group, click Save. When prompted to save all edits, click Yes.**

18. **Save the project, and close the Rematch Addresses and Locate panes.**

19. **Keep the project open, and continue to exercise 7c.**

ON YOUR OWN

You can enter and look up addresses directly in ArcGIS Online. You can also add CSV files to any ArcGIS Online map (up to 1,000 features) as long as it has supported fields, such as latitude, longitude, or address. Records are added to the map as points, which can then be shared with others or exported as data. To learn more about adding CSV files, go to ArcGIS Pro Help > Data > Data types > Tables > Interact with tabular data.

Use geoprocessing tools to analyze vector data

Estimated time to complete: 45 minutes

For this exercise, you will continue using analysis to identify social services that are within a quarter mile, half mile, and one mile of the public housing sites that you geocoded in exercise 7b. Social services are essential to all people living in the community, especially people who are most vulnerable. These programs include early childhood education; food assistance; resources for seniors, foster children, and families living in poverty; cultural enrichment; and after-school child care. Understanding that a goal of public housing sites is to be located near essential social services may provide some insight into whether some areas have lower accessibility to social services.

Exercise workflow

- *Create multiple-ring buffers around public housing sites.*
- *Clip social service sites using public housing buffers.*
- *Perform a spatial join between the buffer and social service sites.*
- *Calculate and measure the distance between public housing sites and social service sites.*
- *Select by location.*
- *Dissolve features.*
- *Intersect layers.*

Create buffers

Buffers are polygons that are created around a feature at specified distances. To look for social services that are within a quarter mile, half mile, and one mile of public housing sites, you'll first create multiple buffers that can be used as a clipping area for the social service sites.

Although the Buffer tool creates *buffers* of only one distance at a time, the Multiple Ring Buffer tool can create concentric buffer rings to represent multiple distances.

1. **Continue working with CommunityHousing.aprx.**

2. **In the Geoprocessing pane, open the Multiple Ring Buffer tool (Analysis Tools > Proximity).** A multiple-ring buffer creates concentric rings around a feature at defined intervals. You will use the tool to create rings with distances of quarter mile, half mile, and one mile around public housing sites. These concentric rings will estimate a straight-line proximity to social services. This measurement is a simplified representation of accessibility. Many more factors help define what accessibility means in real terms: mode of transportation, walkability, and road infrastructure.

3. **In the Input Features drop-down list, choose Public_Housing_Sites.**

4. **For Output Feature class, enter** Public_Housing_Ring_Buffer.

5. **In the Distances box, enter** 0.25, **and press Enter. Also enter distances of** 0.5 **and** 1.0.

6. **In the Buffer Units drop-down list, select Miles.**

7. **In the Dissolve Option drop-down list, make sure that Non-overlapping (rings) is selected.**

8. **Run the tool.**

9. **Apply a graduated color scheme (such as Blue-Green 3 Classes) based on the three distances of the public housing ring buffer.**

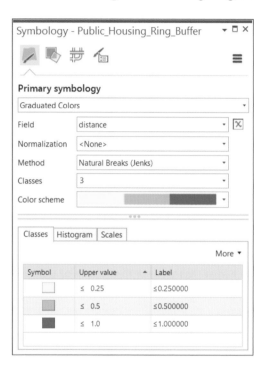

10. **In the Appearance tab, change the Public_Housing_Ring_Buffer layer transparency to 40%.**

11. **Turn on the la_census_tracts layer.**

12. **Zoom into the map to get a better look at the buffers.**

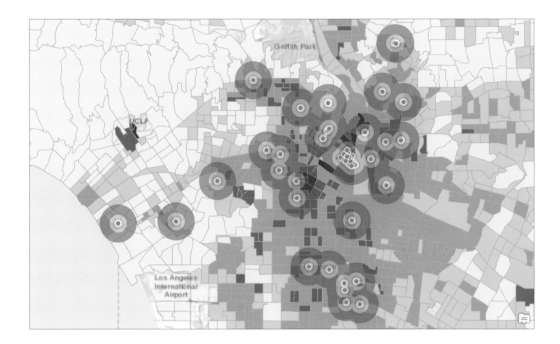

At a glance, you can see a visual cluster in central Los Angeles County. Next, you will add a social services layer to the map. Although many factors may affect how residents of public housing sites use social services, one way to identify how social services are accessed is to determine which ones are nearby. Although the kinds of social services vary from area to area and person to person, understanding which social service sites serve an immediate neighborhood allows government, nonprofit organizations, and other community stakeholders to plan and provide services where necessary.

Clip features

The Clip tool extracts features using other features as a "cookie cutter" or trim boundary. This tool is particularly useful because you want to find only the features that are within a certain distance of the public housing sites.

1. **Turn off the la_census_tracts layer.**

2. **Add the social services layer to the project.**

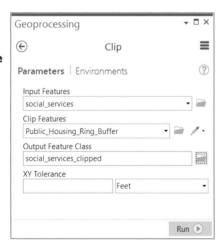

Using this layer gives you an idea of what you are trying to extract—all of the social services that fall within the boundary of the public housing buffer layer.

ON YOUR OWN

View the social services attribute table to see what kinds of organizations serve the communities. Try using the Select tool to manually select social services around public housing sites. When you are finished, clear all selections.

3. **On the Analysis tab, in the Tools gallery, click Clip and set the following parameters. (You may need to click the down arrow to see the next row of quick access tools.)**

   - **For Input Features, choose social_services.**
   - **For Clip Features, choose Public_Housing_Ring_Buffer.**
   - **Enter** social_services_clipped **as the Output Feature Class.**
   - **Run the tool.**

The clipped social services layer is automatically added to the map.

4. **Turn off the social_services layer.**

Now only the social service sites (assigned a random color) that are clipped to the public housing buffer layer are visible.

5. **Open the attribute table of the social_services_clipped layer.**

The *clip* operation reduced the number of social service sites to 1,306 from 5,410. The table has attributes that include name, category, and description of social services. The cat2 attribute includes many categories that can help you if you are focusing on specific services.

6. **After examining the table, close it.**

Create a spatial join

The features in the clipped social services layer are located within a one-mile buffer of public housing sites. However, the social services layer does not have an attribute that indicates which range of the buffer (a quarter mile, a half mile, or one mile) it is in. To include this information, perform a spatial join. Recall from "Exploring geospatial relationships" in chapter 3 that a spatial join essentially joins attributes from one feature to another based on the spatial relationship.

1. **On the Analysis tab, in the Tools gallery, click Spatial Join.**

    Spatial
    Join

2. **In the Geoprocessing pane, for Target Features select social_services_clipped. For Join Features, select Public_Housing_Ring_Buffer. Rename the Output Feature Class to** social_services_buffer_range.

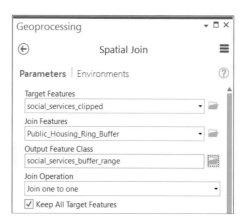

3. **In the Field Map of Join Features section, scroll down the Output Fields list, and choose distance.**
    - **To the right side of the same section, click the Properties category.**
    - **Rename both Field Name and Alias to** range.
    - **Make sure that Type is set to Double (so that the values have a decimal point).**

4. **Near the bottom of the pane, for Match Option select Completely within. Run the tool.**

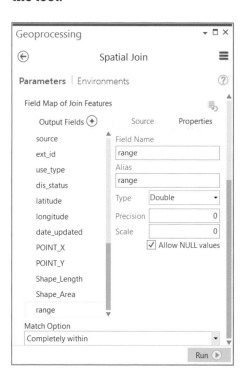

5. **Open the attribute table of the social_services_buffer_range layer.**

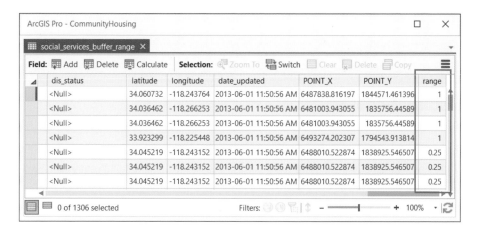

The range field is added to the table so that each social service site now includes the buffer range value of where it is located. This value can be used for a variety of reasons, including selection and visual classification.

ON YOUR OWN

*Change the symbology of the social services buffer range layer so that it includes
three categories—one for each of the buffer ranges. Using a classification can
help represent accessibility to these social service sites. The color scheme
options of your layer may differ, but the general concepts will be the same.*

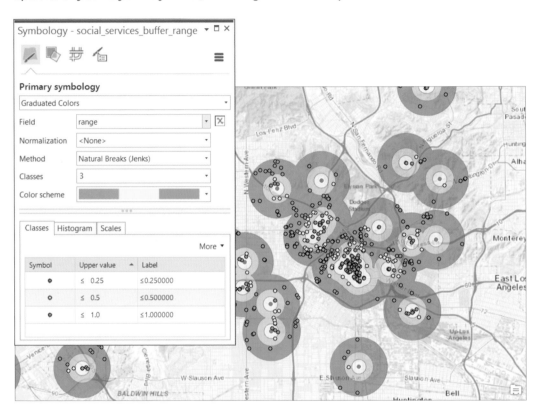

Calculate distances

Although knowing the approximate proximity range is useful, you may want to get an exact
distance between features to know how *near* one is to another. Earlier, you used the Measure
tool to check the distance between features. To add a distance measurement from each social
service site to the nearest public housing site, use the Near tool.

1. **On the Analysis tab, in the Tools gallery, open the Near tool.**

2. **In the Geoprocessing pane,
    for Input Features, choose
    social_services_buffer_range.**

    - **For Near Features, select
      Public_Housing_Sites.**
    - **Enter a Search Radius of** 1,
      and change the units to Miles.
    - **Select the check box for Angle,
      and select Geodesic from the
      Method drop-down list.**
    - **Run the tool.**

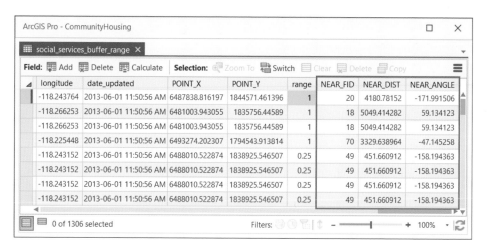

3. **Open the attribute table of the social_services_buffer_range layer.**

longitude	date_updated	POINT_X	POINT_Y	range	NEAR_FID	NEAR_DIST	NEAR_ANGLE
-118.243764	2013-06-01 11:50:56 AM	6487838.816197	1844571.461396	1	20	4180.78152	-171.991506
-118.266253	2013-06-01 11:50:56 AM	6481003.943055	1835756.44589	1	18	5049.414282	59.134123
-118.266253	2013-06-01 11:50:56 AM	6481003.943055	1835756.44589	1	18	5049.414282	59.134123
-118.225448	2013-06-01 11:50:56 AM	6493274.202307	1794543.913814	1	70	3329.638964	-47.145258
-118.243152	2013-06-01 11:50:56 AM	6488010.522874	1838925.546507	0.25	49	451.660912	-158.194363
-118.243152	2013-06-01 11:50:56 AM	6488010.522874	1838925.546507	0.25	49	451.660912	-158.194363
-118.243152	2013-06-01 11:50:56 AM	6488010.522874	1838925.546507	0.25	49	451.660912	-158.194363
-118.243152	2013-06-01 11:50:56 AM	6488010.522874	1838925.546507	0.25	49	451.660912	-158.194363

0 of 1306 selected

In addition to the distance attribute (NEAR_DIST), the Angle option added an angle attribute
(NEAR_ANGLE). The distance values are in feet, which comes from the coordinate system
of the input features. The Geodesic method calculates the shortest distance and takes into
account curvature and distortion of the input layer's coordinate system.

4. **Close any open tables.**

5. On the Edit tab, if necessary, click the Snapping button to turn on feature snapping.

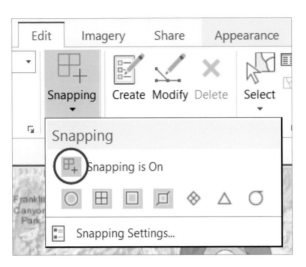

6. On the Map tab, click Measure (make sure it is set to Measure Distance).

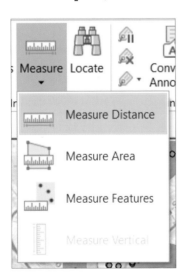

7. Move the mouse cursor over a public housing site. When it snaps to the point, click the point to begin the first line segment.

8. **Drag the mouse away from the public housing site, and click a social service site nearby. Verify the distance from a social service site to the closest public housing site.**

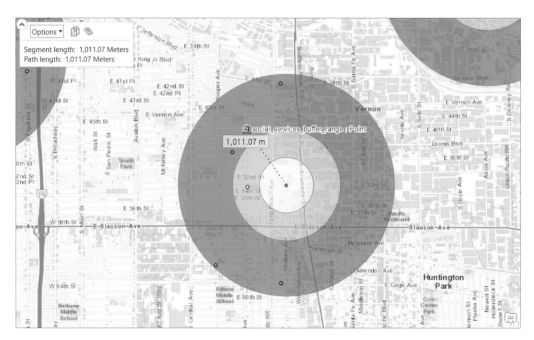

A line segment is drawn temporarily on the map. If you continue to click the map, more segments are drawn. You can also double-click to finish the measurement of a segment.

What is the closest social service site to the Pueblo Del Rio public housing site?

9. **In the upper-left corner of the map, click the Measure Options down arrow, point to Distance Units, and enable Feet.**

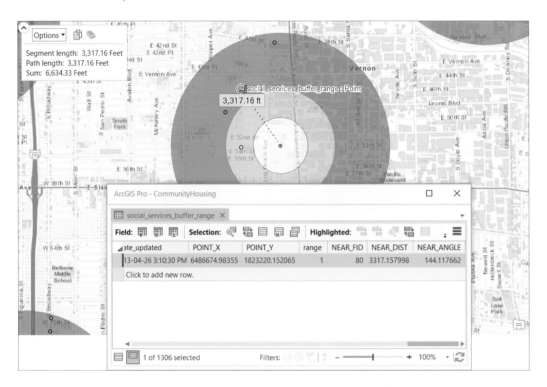

A cumulative sum of the distance or area is also displayed as part of the results.

10. **If you want to cancel a measured line segment, press Esc. Click the Clear Results button to reset all of the measurement results.**

Select by location

To isolate the clusters of census tracts that you have seen throughout this project, you will use the Select By Location tool. This tool can select features by calculating spatial relationships between different layers. This step is also known as a *location query*.

1. **In the Contents pane, turn off any point layers, and turn on the la_census_tracts layer.**

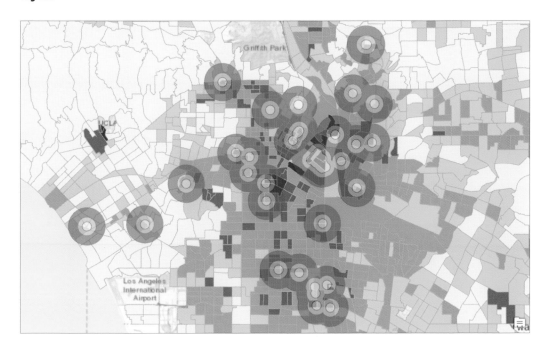

2. **On the Map tab, click the Select By Location button.**

   - **For Input Feature Layer, select la_census_tracts.**
   - **For Relationship, select Have their center in.**
   - **For Selecting Features, select Public_Housing_Ring_Buffer.**
   - **Run the tool.**

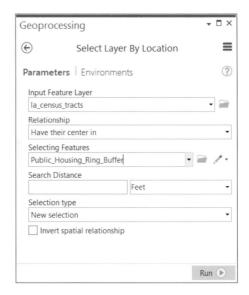

The "Have their center in" relationship means that the center point of a census tract must fall within the buffer area to be selected. This relationship forms a tighter cluster of census tracts around each public housing site, which makes a more relevant sample. To make the selection of clusters its own layer, you will export the clusters to a new layer.

3. **In the Contents pane, select the la_census_tracts layer.**

4. **On the Data tab, click the Export Features button.**

5. **Rename the output feature class to** clusters. **Run the tool.**

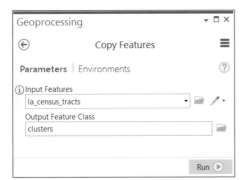

6. **Clear the selection.**

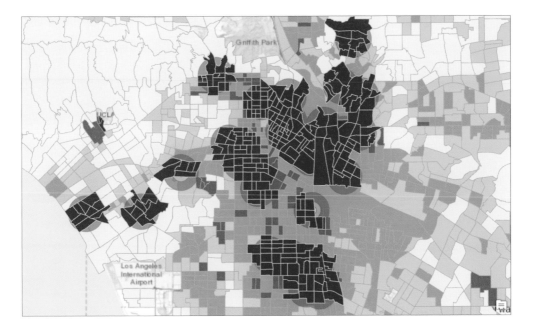

The clusters are now another layer and can be included in other maps or projects. You do not necessarily need them in this map. However, you can also *dissolve* the census tracts in the cluster to make the tract boundaries visible outlines on the map.

Dissolve features

1. **On the Analysis tab, in the Tools gallery, open the Dissolve tool, and modify its parameters.**

   - **For Input Features, select clusters.**
   - **Rename the Output Feature Class** clusters_outline.
   - **Maintain the other default settings, and then run the tool.**

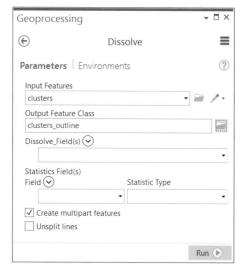

The census tracts are dissolved.

2. **Change the symbology of the clusters_outline so that it has no color and a distinct outline color to stand out from the background.**

3. **Turn off the clusters layer.**

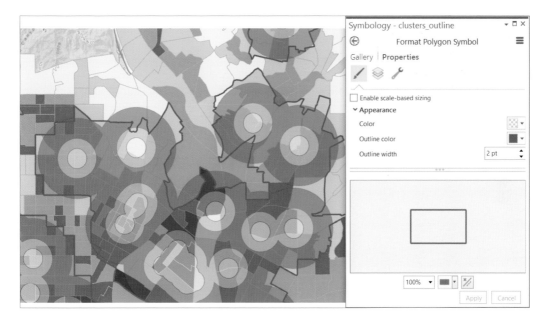

Intersect layers

Before concluding your project, you will add another layer that you can use to finalize your maps if outside agencies or community stakeholders require them. To create this layer, you will run an *intersect* operation on a City of Los Angeles political boundary and the Los Angeles County census tracts layer. The output will contain areas that are coincident to both inputs. Although the results look similar if you use the Clip tool, the difference is that the attributes of both the boundary layer and census tracts layer end up in the output feature class.

1. **Turn off all layers except the basemap, and add the city_boundary layer to the project.**

TIP *Press and hold Ctrl and click any layer's check box to change the visibility of all the layers in the Contents pane.*

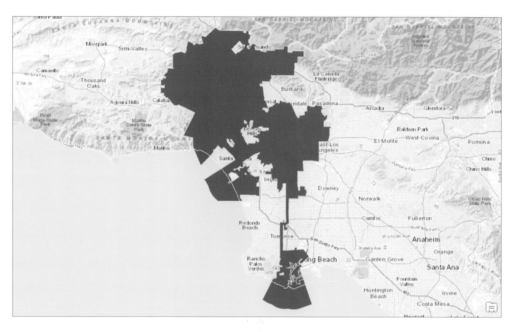

2. **On the Analysis tab, in the Tools gallery, open the Intersect tool and modify its parameters.**

   - **For Input Features, select city_boundary and la_census_tracts.**
   - **Rename the output feature class to** city_census_tracts.
   - **Maintain the other default settings, and run the tool.**

The city_census_tracts layer is added to the map. It retains the Los Angeles city boundary and the census tracts, while retaining all of the attribute information.

3. **Turn off the city_boundary layer.**

4. **Select the city_census_tracts layer, and open the Symbology pane.**

5. **Click the pane options menu, and click Import symbology.**

6. **Choose la_census_tracts as the Symbology Layer.**

7. **In the Symbology Fields section, select VALUE_FIELD as the Type. For both the Source Field and Target Field, select Area Median Income.**

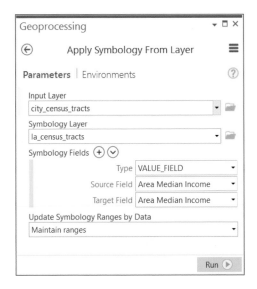

If you see the symbology in the Contents pane but not on the map, in the Symbology pane, set the Field to Average Median Income, and change the Upper value and Label to match the la_census_tracts symbology.

8. **Turn on visibility for the buffers and point layers.**

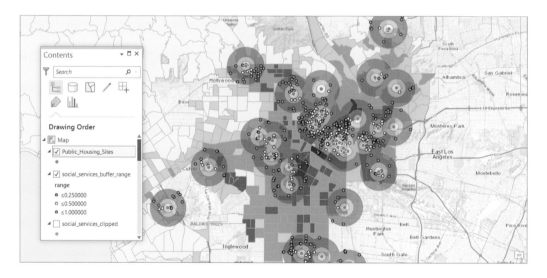

9. **Save the project.**

ON YOUR OWN

Find a cluster of public housing sites, and view the social service categories within the range of public housing buffers. Are these public housing sites being served appropriately by the services around them? Some public housing sites have many services densely located nearby, while others have few.

Try using the Summarize Nearby tool. It finds features that are within a specified distance of features in the input layer and calculates statistics for the nearby features. Distance can be measured as a straight-line distance (similar to what was done in this exercise), a drive-time distance (such as within five minutes), or a drive distance (such as within two miles). Drive-time and drive distance measurements require that you be logged in to an ArcGIS Online organizational account with Network Analysis privileges, and they consume credits.

You are finished with this chapter. If you are not continuing to chapter 8, it is okay to close ArcGIS Pro.

Summary

Using GIS to highlight social services that serve a community is a powerful application for assessing community well-being within a city or region. With layers of information about local geography, roads, building sites, socioeconomic conditions, and more, GIS can provide planners and the community solutions to the challenges that residents of public housing face in accessing social services. Planners can use GIS to map household income data as an index to advocate for more affordable housing. Policy-makers can use GIS to identify development opportunities for housing, commercial, and social services that are compatible with the neighborhood infrastructure needs of mixed-income communities. Although much research and analysis has been done, the effectiveness of social assistance and public housing continues to warrant further analysis.

Glossary terms

geocoding
address locator
overlay
buffer
clip
near
location query
dissolve
intersect

CHAPTER 8
Analyzing spatial and temporal patterns

Exercise objectives

8a: Create a kernel density map

- Select by attributes
- Create a kernel density

8b: Perform a hot spot analysis

- Run the Optimized Hot Spot Analysis tool
- Create a space-time cube
- Visualize a space-time cube
- Run the Emerging Hot Spot Analysis tool

8c: Explore the results in 3D

- Switch to a local view
- Change 3D visualization styling

8d: Animate the data

- Enable time
- Animate using the Time slider
- Animate using the Range slider

Incidents of crime are neither uniform nor randomly organized in space and time. Crime mapping analysts can use GIS to identify spatial patterns to gain a better understanding of the role of location, proximity, and opportunity, while providing key decision-makers with information to put crime prevention solutions in place.

Time is an important variable in analyzing crime. When using temporal data—in this case, crime incidents collected at a specific location and specific time—it is possible to see how patterns of crime shift over time and geography. As long as data is collected or surveyed to represent an instant in time—over the course of one day, one month, one year—the inclusion of time in a dataset's attributes is what gives temporal data the necessary information for a GIS to perform spatial and temporal analysis. The development of crime mapping has introduced the importance of location and time in criminology.

Scenario: you work as a crime analyst for the City of Philadelphia, and your role is to use GIS to examine criminal activity over time. You will do this by identifying crime hot spots and locating where those hot spots have gotten worse over time to help create spatially targeted crime prevention plans. Using a list of crime incidents that include *temporal data*—that is, data that has a time attribute—you will isolate a specific type of crime and create map layers that show crime locations as well as create crime hot spots.

GIS IN THE WORLD: SMART CRIME FIGHTING

The Ogden Police Department (OPD) Real Time Crime Center in Ogden, Utah, provides 24/7 support to law enforcement and uses ArcGIS to automatically link crime and other datasets maintained in different databases. OPD can perform advanced analysis and digitally map the results. These functions have allowed police staff to effectively deter crime and make arrests. Read more about the project here, "Better Policing through Analysis": www.esri.com/news/arcnews/spring12articles/better-policing-through-analysis.html.

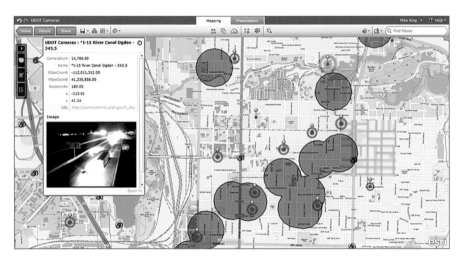

The Ogden Police Department relies on ArcGIS® Server, as well as ArcGIS.com, to share information with the public and partnering police agencies. *Pictured:* known burglary suspects are shown inside a 1,000-foot buffer of forced-entry residential burglaries. Courtesy of Ogden (Utah) Police Department.

DATA

In the \\EsriPress\GTKAGPro\CrimeIncidents\CrimeIncidents geodatabase:

- crime—a feature class that represents crime incidents for the year 2014 in the City of Philadelphia (*Source: City of Philadelphia Police Department*)

EXERCISE 8A

Create a kernel density map

Estimated time to complete: 20 minutes

In this exercise, you will explore the crime layer of crime incidents and look more closely at its attributes. You will then create a kernel density layer to approximate crime hot spots to see where they are located in the city.

Exercise workflow

- *Review data to see what makes it temporal.*
- *Select robberies by attribute, and create a layer from the selection.*
- *Create a kernel density layer to represent a crime hot spot.*

Select by attributes

1. **In ArcGIS Pro, open CrimeIncidents.aprx from your \\EsriPress\GTKArcGISPro\ CrimeIncidents folder.**

2. **Open the crime layer attribute table.**

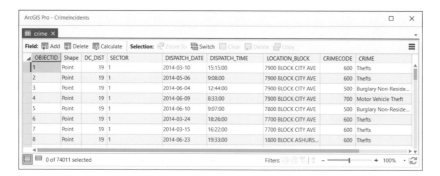

The crime layer contains crimes for the 2014 calendar year in the City of Philadelphia. The table lists 74,011 crime incidents, which are represented at a per block group address level, not an actual address. This layer of anonymity protects the privacy of those involved with each incident. In addition, names and other identifiable information are not included with each incident. Notice the dispatch date and dispatch time attributes for each record. Although it is entirely possible to create a day-by-day temporal map with this level of data granularity, you will instead create a series of temporal maps for each month and use it in a time animation. In addition, you will focus on one type of crime: robberies—incidents with and without a firearm. To perform this analysis, you will first extract data by type of crime.

3. **Close the attribute table.**

4. **On the Map tab, in the Selection group, click Select By Attributes.**

5. **In the Geoprocessing pane, select crime from the Layer Name or Table View drop-down list. Select New selection from the Selection type drop-down list.**

6. **Under Expression, click Add Clause.**

7. **Select CRIME from the first drop-down list, make sure "is Equal to" is selected in the operator drop-down list, and select Robbery Firearm in the third drop-down list. Click Add.**

8. **Click Add Clause again. Select Or from the first drop-down list, and build the rest of the expression to select Robbery No Firearm.** The expression includes the "or" *operator*, which selects both robbery attributes—with firearms and without firearms—and includes them in the new selection.

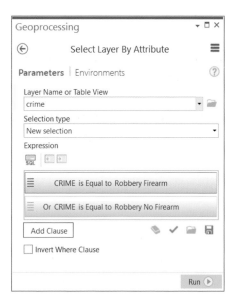

9. **Run the tool.** Some 7,006 robbery points are selected on the map. Next, you will create a feature layer from the selection, in which you will work to extract robbery crimes by month.

10. **Right-click the crime layer, and go to Data > Export Features.**

11. **In the Geoprocessing pane, select crime from the Input Features drop-down list.**

12. **Point to the Output Feature Class box to see the default output location—the project geodatabase. Maintain this path, but change the Output Feature Class name to robbery.**

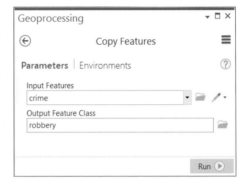

13. **Run the tool.** The robbery layer is now added to the map, and you can now select robberies by month.

14. **Clear the selection, and turn off the crime layer.**

15. **Create a selection by attribute in which DISPATCH_DATE is on or after 1/1/2014 and DISPATCH_DATE is on or before 1/31/2014. Run the tool.** Some 640 robbery incidents during the month of January are selected on the map. Similarly, you will create a layer from this selection.

16. Export the selected features, and name the layer robbery_jan.

17. Clear the selection, and hide the robbery layer.

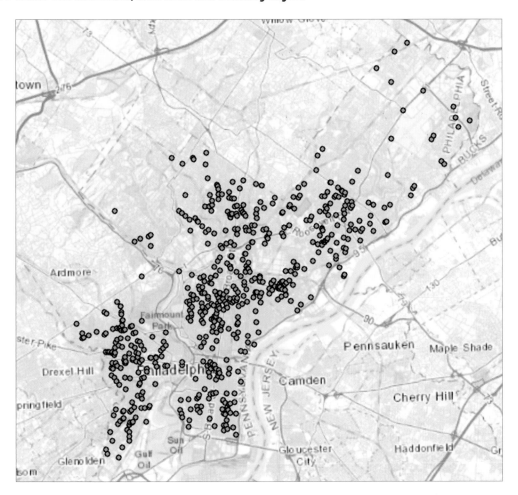

Next, you will create a kernel density for the robberies in January.

Create a kernel density

A *kernel density* calculates the density of features in an area around those features and is one of the most common techniques in crime mapping.

1. **On the Analysis tab, in the Tools group, click Kernel Density.**
    Kernel
    Density

2. **In the Geoprocessing pane, set the following kernel density parameters:**
    - **For Input point feature, select robbery_jan.**
    - **For Population field, make sure NONE is selected.**
    - **For Output raster, replace the name with** robbery_jan_density.
    - **If necessary, change Area units to Square miles.**

    The planar method is used because the map's coordinate system is WGS84 Web Mercator Auxiliary Sphere, and the robbery data layer is stored in a geodetic coordinate system (WGS84).

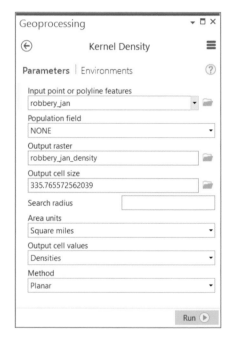

3. **Run the tool.** The kernel density is calculated and added to the map. The density is automatically classified with 10 classes and styled using a random color ramp (your colors may look different from the graphic).

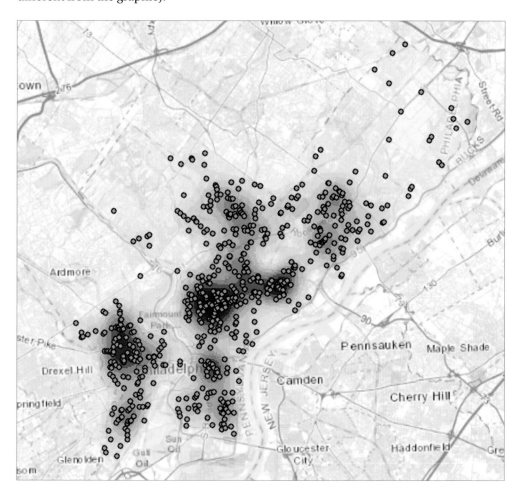

4. **On the Appearance tab, change the robbery_jan_density layer transparency to 40%.** The least dense class of the kernel density now has no color, and the remaining classes at 40 percent transparency allow you to see the basemap.

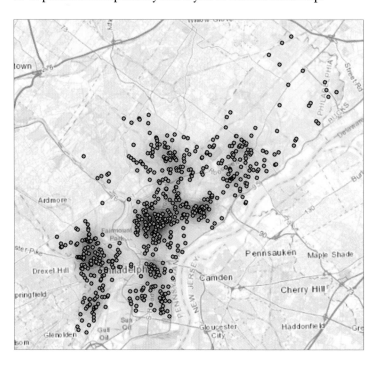

Next, you will use more analysis tools on the robbery layer to help visualize more of the hot spots.

ON YOUR OWN

Try creating a kernel density for other months to compare how each month differs. Try creating a kernel density for other categories of crime. Do some areas have more incidents of certain crimes than others? It is important to note when comparing density surfaces that the classifications are the same. Using different classifications causes different results.

5. **Save the project.**

EXERCISE 8B

Perform a hot spot analysis

Estimated time to complete: 45 minutes

You created a kernel density in exercise 8a to briefly see where areas of high density may occur. You will expand on how to build out an analysis by using other types of analysis tools. In this exercise, you will run an optimized hot spot analysis tool to find statistically significant hot spots and cold spots of robberies. You will also create a space-time cube, a data pattern mining tool, to look for trends over time.

Exercise workflow

- *Run the Optimized Hot Spot Analysis tool to create a hot spot map.*
- *Run the Create Space Time Cube tool to mine for data patterns.*
- *Visualize the space-time cube in 3D.*
- *Run the Emerging Hot Spot Analysis tool to see hot spot trends in the robbery data.*

Run the Optimized Hot Spot Analysis tool

1. **Continue working from the map in exercise 8a.**

2. In the Contents pane, hide the **robbery_jan** layer and the **robbery_jan_density** layer. Turn on the **robbery** layer. (If you did not create these layers, add them from the geodatabase in the Results folder.)

3. On the Analysis tab, scroll or expand the Tools, and click **Optimized Hot Spot Analysis.**

    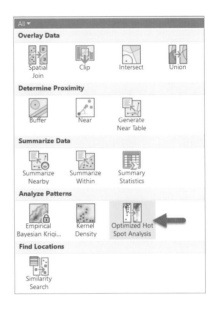

4. In the Geoprocessing pane, for Input Features, select **robbery**. Rename the Output Features to robbery_OHSA. **Leave the other settings as is.**

    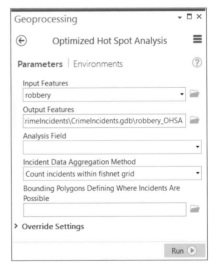

5. Run the tool.

6. Hide the robbery layer.

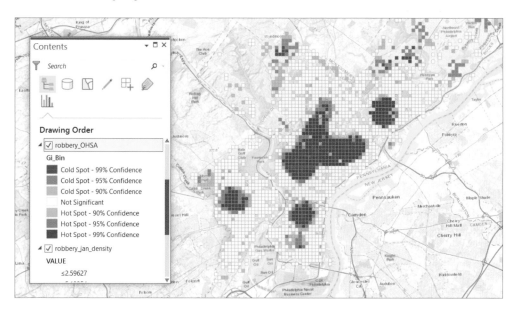

The optimized *hot spot analysis* performed on the robbery data is added to the map. The tool aggregated robbery incidents into weighted features and analyzed their distribution to identify an appropriate scale of analysis. It then determined which areas are significant, including areas that are hot (large clusters symbolized in red) and cold (smaller clusters symbolized in blue). Areas that are not significant are symbolized in yellow.

7. Open the robbery_OHSA attribute table.

OBJECTID	Shape	SOURCE_ID	JOIN_COUNT	Shape_Length	Shape_Area	GiZScore Fixed 3987	GiPValue Fixed 3987	NNeighbors Fixed 3987	Gi_Bin Fixed 3987_FDR
1	Polygon	1	1	3716.000442	863041.205308	-1.866372	0.061989	7	0
2	Polygon	2	2	3716.000442	863041.205294	-1.813375	0.069774	6	0
3	Polygon	3	2	3715.999786	863040.900509	-1.45257	0.146343	6	0
4	Polygon	4	1	3715.999786	863040.900523	-2.032668	0.042086	10	0
5	Polygon	5	1	3715.99913	863040.595724	-1.754995	0.07926	7	0
6	Polygon	6	1	3715.999786	863040.900509	-2.176914	0.029487	11	-1
7	Polygon	7	1	3716.000442	863041.205308	-1.366213	0.171872	3	0
8	Polygon	8	1	3716.000442	863041.205308	-2.125926	0.033509	10	-1
9	Polygon	9	3	3715.99913	863040.595724	-2.176914	0.029487	11	-1

ArcGIS Pro - CrimeIncidents

robbery_OHSA

Field: Add Delete Calculate   Selection: Zoom To Switch Clear Delete Copy

0 of 1904 selected   Filters: + 100%

Before you look at the values, it is important to make some assumptions used in statistics. You are essentially asking the questions: "Is this pattern the result of random spatial processes?" and "How likely is it that this pattern is random?" The initial assumption in this cluster analysis is that crime clusters happen in specific areas of the city and are not random. When it is *very* unlikely that the pattern is random, you reject the null hypothesis—which, in this case, is complete spatial randomness—and you can feel confident that the patterns you are seeing are evidence of underlying spatial processes. In other words, you have found statistically significant clusters of both high and low values that you can trust, and you can begin to ask questions as to why they occur, thereby rejecting the null hypothesis.

The intensity of robbery clustering is determined by the z-score returned in the output. Z-scores are standard deviations and are represented in the table as the attribute named GiZscore Fixed 3997. As an example, a z-score of +2.5 means that the result is 2.5 standard deviations and is at a high confidence level (greater than 95 percent), typically indicating statistically significant clustering.

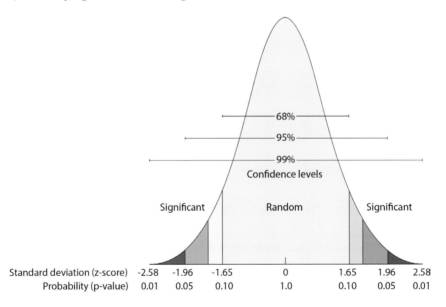

| Standard deviation (z-score) | -2.58 | -1.96 | -1.65 | 0 | 1.65 | 1.96 | 2.58 |
| Probability (p-value) | 0.01 | 0.05 | 0.10 | 1.0 | 0.10 | 0.05 | 0.01 |

A probability graph shows confidence levels and standard deviation and probability values that indicate either the significance or the randomness of phenomena.

8. **Use the Explore tool to select one cell in a red hot spot.**

Most likely, the z-score of the selected cell is greater than +2.5. However, the p-value is a very low number, close to zero. The p-value represents the probability that the observed spatial pattern was created randomly. A very small p-value means that it is very improbable that the observed spatial pattern is the result of random processes. This small p-value can be an indication that robbery incidents are not entirely random and that many contributing factors can lead to their occurrence, such as socioeconomic factors in areas of low-income neighborhoods.

9. **Close the attribute table, and clear the selection.** Next, you will create a space-time cube to add a temporal component to the analysis.

Create a space-time cube

The Create Space Time Cube tool summarizes the set of robbery incidents into a data structure by aggregating them into space-time bins. These bins can be viewed in either two or three dimensions. Each bin contains the robbery points for a certain month, and the pattern of incidents over time is analyzed for all locations. Creating a *space-time cube* can help you aggregate the robbery points, and visualizing it will help you understand how the robbery data is distributed over a geographic area. This tool only creates the space-time cube—it does not visualize it. You will visualize the space-time cube immediately after creating it.

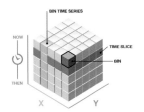

A set of points can be summarized into a data structure by aggregating them into space-time bins that form a cube.

1. **In the Geoprocessing pane, search for** create space time cube. **In the list of search results, click Create Space Time Cube by Aggregating Points.**

2. **For Input Features, select robbery. Rename Output Space Time Cube to** robbery_STC.nc. **(Hint: it is important to include the .nc file extension.)** The .nc file extension stands for the netCDF format (network Common Data Form) and is primarily used for storing multidimensional data such as temperature, but it can also be used for other data types. ArcGIS Pro uses the capabilities of the netCDF format and can display data in multiple dimensions, including time.

3. **For the time field, select DISPATCH_ DATE, and change the following parameters:**
    - **In the Time Step Interval box, enter** 1.
    - **From the drop-down list, select Months.**
    - **Run the tool.**

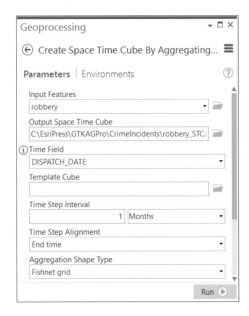

4. **When the tool is finished running, click the message to see more details about what was created.**

The message provides insight into what was created, including the number of locations, the geographic size of each location, the total geographic size, and the number of time intervals.

You will use another tool to visualize the space-time cube.

Visualize a space-time cube

The space-time cube you created for robbery incidents will be rendered in 3D, so you will create a new 3D scene.

1. **On the Insert tab, click the New Map down arrow, and then click New Local Scene.**

2. **In the Contents pane, right-click Scene, and click Properties.**

3. **Go to the Elevation Service tab, expand Ground, and then remove the elevation services that are being used by default. Click OK.**

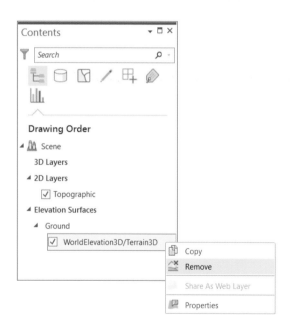

Instead of elevation, time is being used as the vertical axis in visualizing the robbery space-time cube. By removing the service that provides elevation data, all of the time values will start on the ground and at the same base, with an elevation equal to zero.

4. **In the Geoprocessing pane, search for** visualize space time cube. **In the list of search results, click Visualize the Space Time Cube in 3D.**

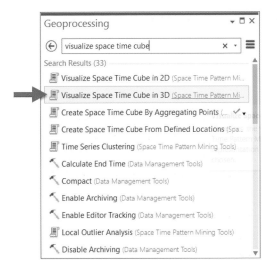

5. **In the Geoprocessing pane, browse to the folder location where you saved the robbery_STC.nc file, and change these parameters:**
   - **For Cube Variable, select COUNT.**
   - **For Display Theme, select Value.**
   - **For the Output Space Time Features name, enter** robbery_STC_3D.
   - **Run the tool.**

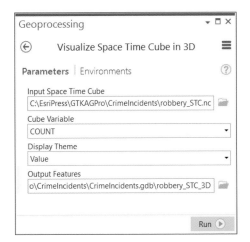

The Count display variable shows the density of the robbery counts in each space-time bin. After the tool is finished running, it may take a few moments to render the scene.

6. **On the Map tab, click Explore, and use the mouse to navigate the map. Zoom in or use Zoom To Layer to see the space-time cubes.**

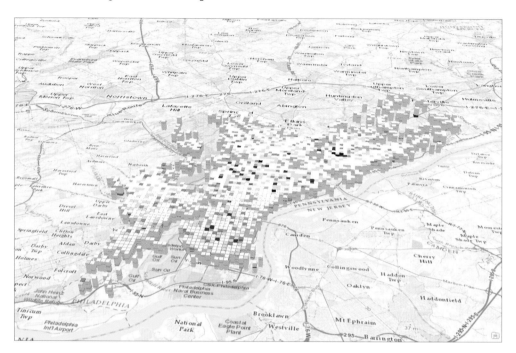

The map shows space-time bins and the counts for robbery incidents. When zoomed in close enough to the scene, you can see 12 intervals, each representing a month, for each location (as shown in the next image).

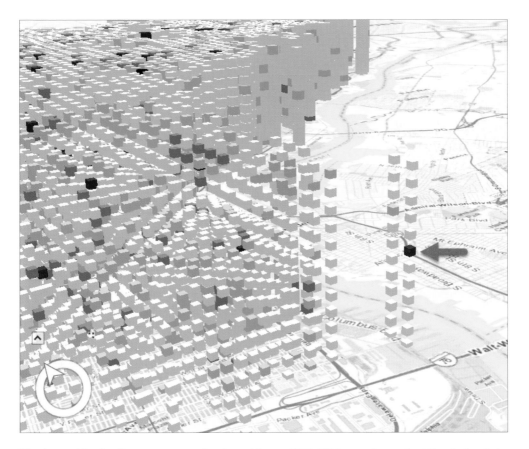

The legend in the Contents pane for the robbery_STC_3D layer shows that the darker intervals show higher counts. Although looking at the space-time cube alone provides some hints on the clustering of robberies, it is not conclusive. You will use another tool that specifically identifies trends in the clustering of space-time point data for a more detailed perspective.

Run the Emerging Hot Spot Analysis tool

The Emerging Hot Spot Analysis tool identifies hot spot and cold spot trends in your data. It essentially provides a similar hot spot analysis at the start of this exercise.

1. **Return to the 2D map, and turn off the robbery_OHSA layer.**

2. **In the Geoprocessing pane, search for** emerging hot spot analysis. **In the list of search results, click Emerging Hot Spot Analysis.**

3. **In the Geoprocessing pane, browse to the folder location where you saved the robbery_STC.nc file.**

4. **For Analysis Variable, select COUNT. Enter** robbery_STC_EHSA **as the Output Features name.**

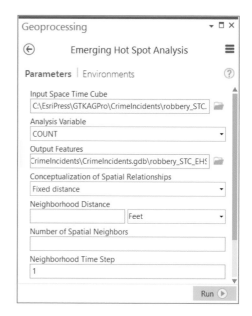

5. Run the tool.

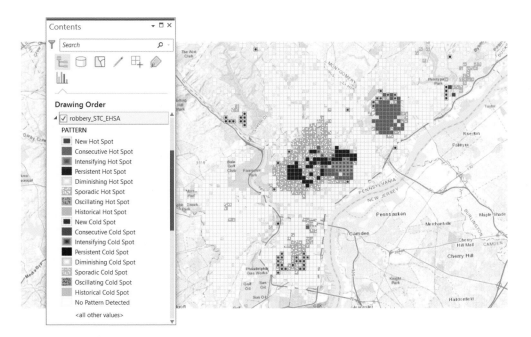

When the analysis is complete, an emerging hot spot analysis layer is added to the map. In the Contents pane, notice the many patterns listed to categorize the locations. Remember that the Emerging Hot Spots Analysis tool takes into account space and time and analyzes all robbery incidents to determine whether clustering is statistically significant.

6. Zoom in to the large hot spot at the center of the map.

7. **Click a location that is classified as Persistent Hot Spot.**

In this classification, a persistent hot spot is a location that has been a statistically significant hot spot for 100 percent of the time-step intervals with no discernible trend indicating an increase or decrease in the intensity of clustering over time. This hot spot classification means that robberies in this area are frequent and concentrated.

8. **Click a location that is classified as Intensifying Hot Spot.** An intensifying hot spot is a location that has been a statistically significant hot spot for 90 percent of the time-step intervals. It is an area in which there has been an increase in the intensity of clustering over time and is statistically significant. The spatial correlation between intensifying hot spots and persistent hot spots is not random. Crime can have a spreading trend, which may be what is happening here.

9. **Pan the map south, and click a location that is classified as New Hot Spot.**

This area has many locations classified as new hot spots. A new hot spot is a location that is a statistically significant hot spot for the final time step and has never been a statistically significant hot spot before. Although many factors can cause this area to become a significant cluster, the map can be used to help determine what is causing the increase in robberies during that time interval.

TIP *To learn more about the Emerging Hot Spot Analysis patterns, go to ArcGIS Pro Tool Reference > Tools > Space Time Pattern Mining toolbox > Space Time Pattern Mining toolbox > Emerging Hot Spot Analysis.*

10. **Clear any selections you have made.**

11. **Make Scene the active map.**

12. **Run the Visualize the Space Time Cube in 3D tool again.**

13. **This time for Display Variable, select "Hot and cold spot results."**

14. Rename the output to robbery_STC_3D_Hotspots.

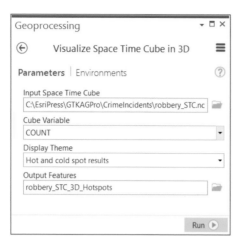

15. Run the tool.

16. Turn off the robbery_STC_3D layer.

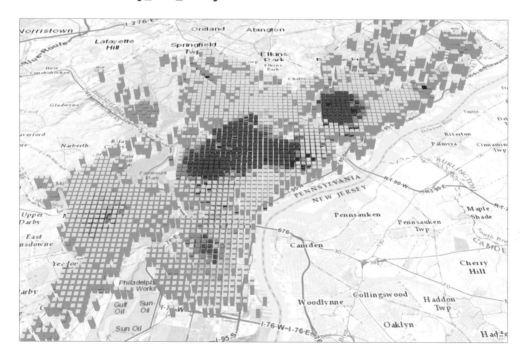

A 3D version of the results of the Emerging Hot Spot Analysis is added to the Scene map.

ON YOUR OWN

Dock the maps horizontally or vertically, and link views to explore both the 2D and 3D hot spot and cold spot analysis maps at the same time.

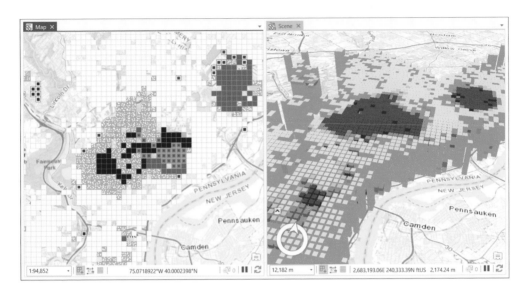

17. **Save the project.**

EXERCISE 8C

Explore the results in 3D

Estimated time to complete: 30 minutes

The ability to animate data is a great visualization tool to show changes over time. In exercises 8a and 8b, you completed the creation of crime hot spots for the City of Philadelphia. In this exercise, you will explore analytical results in three dimensions to observe patterns or trends that may emerge as time passes.

Exercise workflow

- *Switch to a more scientific visualization display.*
- *Investigate hot spots and change visualization styling.*

Switch to a local view

Depending on your ArcGIS Pro settings, the 3D view can be in global viewing mode, which means that the data is being reprojected to WGS84. For this dataset, with a projected coordinate system and a limited area of interest, you should work directly in its native coordinate system, NAD 1983 State Plane Pennsylvania South.

1. **Continue working from the map in exercise 8b. If you linked your views and have them docked horizontally or vertically, you can continue that way. Otherwise, unlink the views, and click Scene to make it the active map.**

2. **Turn off the robbery_STC_3D_Hotspots layer, and make the robbery_STC_3D layer visible.**

3. **On the View tab, click Local.**

4. **In the Contents pane, right-click Scene, and click Properties.**

5. **Click the Coordinate Systems tab to view the coordinate system details.**

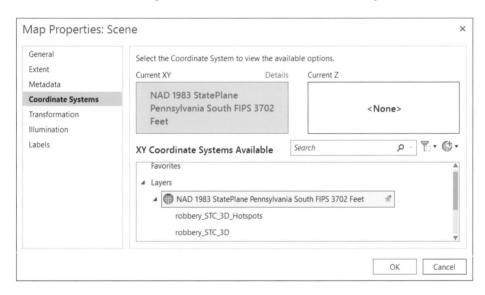

The local view is used for areas of smaller geographic extent in a projected coordinate system or when representing the earth's curvature is not required. In this case, the local view is using the NAD 1983 StatePlane Pennsylvania South FIPS 3702 Feet coordinate system. Next, you will clip the extent.

6. **Click Cancel to close the Map Properties dialog box.**

7. **On the Extent tab, click the Custom extent option.**

8. **From the Calculate from drop-down list, select robbery_STC_3D (make sure Features in current extent only is unchecked), and select the "Clip layers to extent" check box.**

Map Properties: Scene   ✕

General	○ Extent of data in all layers
Extent	◉ Custom extent
Metadata	Calculate from robbery_STC_3D ▾
Coordinate Systems	☐ Features in current extent only
Transformation	
Illumination	Units Spatial Reference (US Feet) ▾
Labels	

                                          Top
                                    303,132.48 ftUS ⬍

                Left   Right
          2,663,015.47 ftUS ⬍   2,747,554.47 ftUS ⬍

                                        Bottom
                                    211,161.48 ftUS ⬍

          ☑ Clip layers to extent

                                                        OK   Cancel

The custom extent is calculated from the robbery_STC_3D layer. Clipping the extent of the view will help isolate the area of interest.

9. **Click OK.**

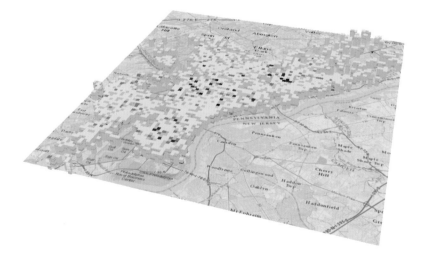

The map properties are applied, and the scene view is updated. You will add a new field so that you can assign it the appropriate time.

10. **Open the attribute table of the robbery_STC_3D layer, go to the pane options menu, and click Fields View.**

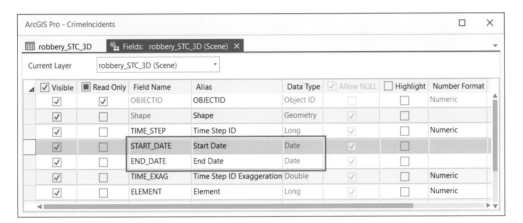

The robbery hot spots are all designated with time. Next, you will enable time so that the hot spots can be included in animations.

11. **Close the Fields View and the attribute table.**

Change 3D visualization styling

Using the standard or default symbology, the cubes will "separate" as you get closer. This type of visualization is fine for some use cases—for example, it helps with column-based understanding of the data and helps to reveal any patterns that might be "wrapped inside" other items.

1. **Hide the robbery_STC_3D layer, and make the robbery_STC_3D_Hotspots layer visible.**

2. **Zoom in until you can see the time intervals separate.**

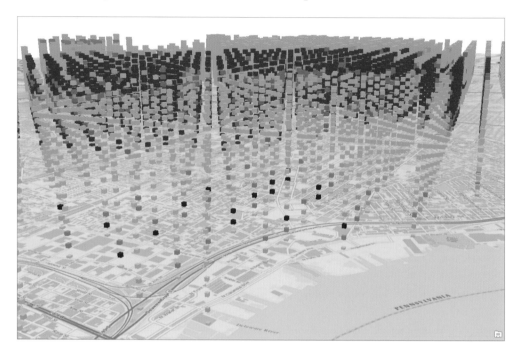

Although this type of visualization works in some cases, in this mode it is harder to understand the spatial coverage of the analysis. In this case, you can resymbolize the content so that the cubes are in fixed real-world size, as flattened cubes that cover the spatial extent they represent.

3. **Open the layer properties of the robbery_STC_3D_Hotspots layer.**

4. Go to the Display tab, and select the "Display 3D symbols in real-world units" check box. Click OK.

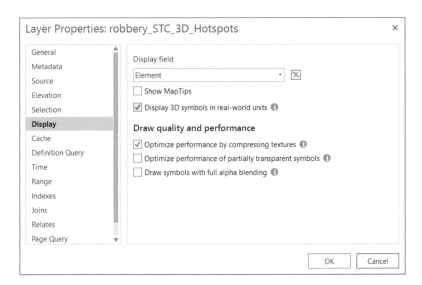

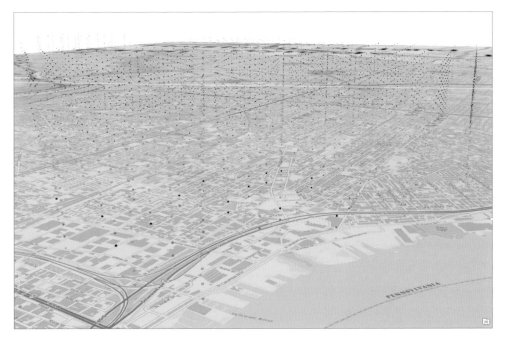

3D symbols displayed in real-world units display a constant measurable size, regardless of view distance. Next, change the symbology of the locations to make them real-world size so that they fit the physical space they represent. You can make this change by manually changing each symbol, using attribute-driven symbol properties (if available on the feature class), or you can use the attribute-driven property settings with constant values. You will use this third option because it is easy to access from the Symbology pane and will not change the legend in the Contents pane.

5. **Right-click the robbery_STC_3D_Hotspots layer, and click Symbology.**

6. **In the Symbology pane, click the "Vary symbology by attribute" button.**

7. **Expand the Size setting, and adjust the following properties:**
    - **Uncheck the Maintain aspect ratio check box.**
    - **Next to the Height field, click the Expression Builder button.**

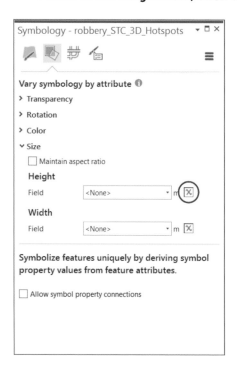

8. **In the Expression Builder window, enter** 50 **in the text box, and then click OK.**

9. **For Width, use the Expression Builder and enter** 250 **in the text box, and then click OK.**

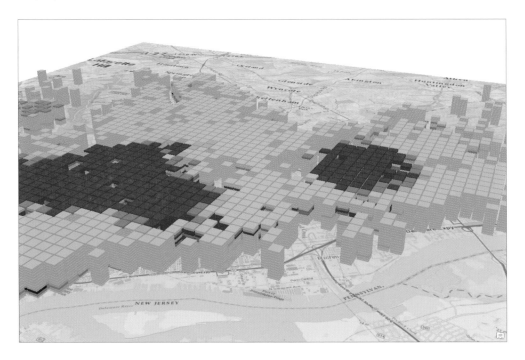

The hot spots of the space-time cube adjust according to the values you entered in the Size property of the Symbology pane. Next, you will adjust the colors to emphasize the significant hot spots. Using transparency, you can hide the nonsignificant locations by making them transparent.

10. **In the Symbology pane, click the Primary symbology button to return to the main symbology settings for robbery_STC_3D_Hotspots.**

11. In the Values section, click the symbol for Not Significant.

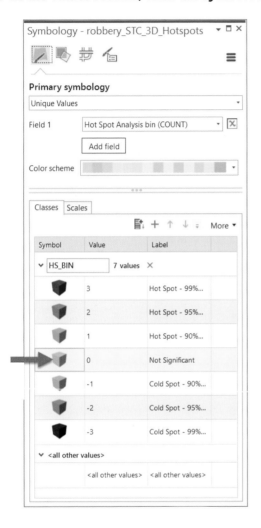

12. **Click Properties. Click the Color down arrow, and select No color. Then click Apply.**

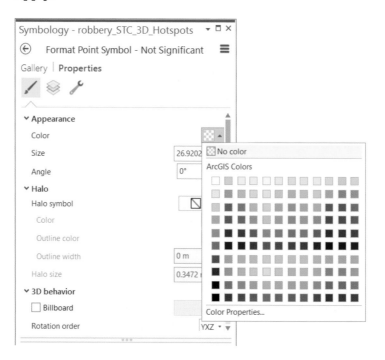

For the hot spots that are designated as 90% Confidence, you will make them partially transparent and therefore less prominent.

13. **Click the back arrow, and then click the symbol for Hot Spot – 90% Confidence.**

14. **Click the Color down arrow and select Color Properties. In the Color Editor, change the transparency to 40%, and then click OK.**

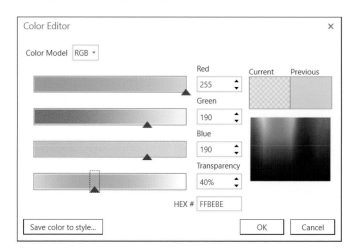

15. **Click Apply.** The map is updated to reflect the changes in the Symbology pane.

ON YOUR OWN

If you are having trouble seeing the hot spots, change the visibility of the robbery incidents layer—by either turning it off or adjusting its transparency. You can also enable the Swipe tool to reveal layers beneath the chosen layer.

16. **Save the project.**

Animate the data

Estimated time to complete: 20 minutes

Exercise workflow

- *Enable time for the 3D hot spot layer.*
- *Set up the Time slider and animate the data.*
- *Set up the Range slider to filter the data.*

Enable time

The robbery incidents layer includes a DISPATCH_DATE attribute that you can use to indicate time. First, you will enable the time for the 3D robbery hot spot layer.

1. **Continue working from the map in exercise 8c. Make sure Scene is the active map.**

2. **Open the properties of the robbery_STC_3D_Hotspots layer.**

3. **Go to the Time tab, and adjust these properties:**
   - **Select "Each feature has start and end time fields" from the Layer Time drop-down list.**
   - **Select Start Date from the Start Time Field drop-down list.**
   - **Select End Date from the End Time Field drop-down list.**
   - **If the Time Extent does not refresh automatically, click the Calculate button, and then click OK.**

Layer Properties: robbery_STC_3D_Hotspots   ✕

General	
Metadata	Layer Time Each feature has start and end time fields ▾
Source	Start Time Field Start Date ▾
Elevation	End Time Field End Date ▾
Selection	
Display	
Cache	Time Extent 2013-12-31 12:00:01 AM 📅 - 2014-12-31 📅
Definition Query	Calculate
Time	☐ Data is a live feed
Range	Rate 1 Seconds ▾
Indexes	
Joins	Time Zone <None> ▾
Relates	☐ Adjust For Daylight Saving
Page Query	Learn more about time properties

OK   Cancel

The Calculate button ensures that the full time extent—start time and end time—is specified. The extent can also be changed later when adjusting the time settings.

You may notice a bar appear near the top of the map or a small box in the upper-right corner of the map that says Time. This Time slider indicates that the map has successfully detected the temporal data that you specified.

Animate using the Time slider

1. **Locate the animation controls in the upper-right corner of the map frame. If necessary, click the Time down arrow to expand the animation controls or point to the animation controls to see them.**

2. **Click the Time tab to make it active. (If the start or end dates are not showing, on the Time slider, click the Enable time button.)**

The Time tab contains all of the settings to control animations. The time extent is indicated in the Full Extent group showing a Start time of 12/31/2013 and an End time of 12/31/2014. Because the time extent shows a full 12 months, all of the robbery incidents displayed represent a full year and are shown on the map. You will adjust the start time slightly.

3. **In the Full Extent Group, change the Start date to 2014-01-01.**

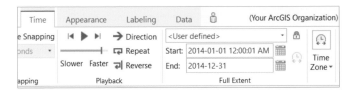

Next, you will adjust the current time to show only the month of January 2014.

4. **In the Current Time group, change the Start date to 2014-01-01 and the End date to 2014-01-31.**
Setting the current time creates a map time that represents the month of January. On the map, notice that the number of robbery hot spots is reduced. These robberies are the ones that occurred only in the month of January. The robbery points for other months are hidden until the map time is changed.

5. **In the Step group, clear the "Use time span" check box. The Step Interval option becomes enabled. Specify an interval of 1 Months.**

6. **Click the Step Forward button.** The map time moves forward one month, and the robbery hot spots for February are shown. On the Time slider and in the Current Time group, the February start and end times are displayed (2014-02-01 to 2014-02-28).

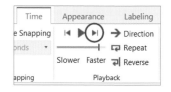

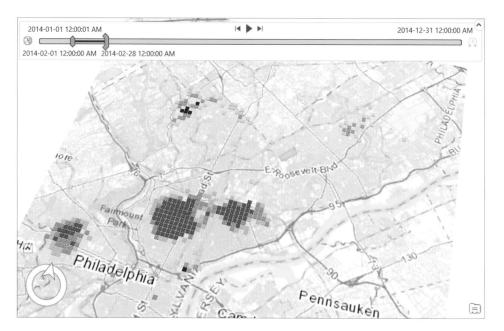

7. **In the Playback group, use the Speed slider to adjust the speed of the animation. Click Repeat to enable it.**

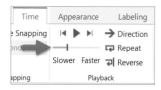

8. **Click the Play button.** The animation begins to play and cycle through each month automatically.

9. **Click the Pause button at any time to stop the animation. Use the Step Back and Step Forward buttons to move to the previous or next interval.** These controls allow you to move map time month by month so that you can see the difference elapse between each month.

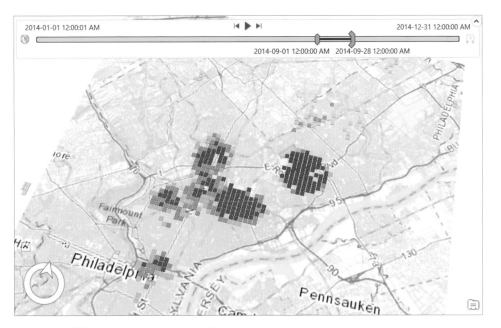

September 2014 robbery hot spots in Philadelphia.

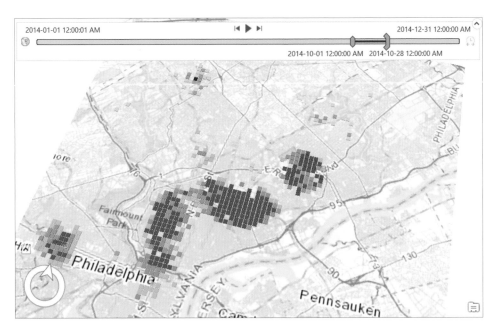

October 2014 robbery hot spots in Philadelphia.

TIP *Unique temporal slices are included in the visualization cache so that the animation displays and performs better on second viewing.*

ON YOUR OWN

Experiment with the Time slider, and adjust the playback options.

Animate using the Range slider

First, you will enable the range of values from an attribute. In this case, you will use the Hot Spot Analysis Bin (Count) attribute, which is the confidence level indicated in the legend. It ranges from −3 to +3, with zero classified as Not Significant.

1. **Open the properties of the robbery_STC_3D_Hotspots layer.**

2. **Go to the Range tab, and click the Add Range button.**

3. **From the Field drop-down list, select Hot Spot Analysis bin (COUNT), and then click Add.**

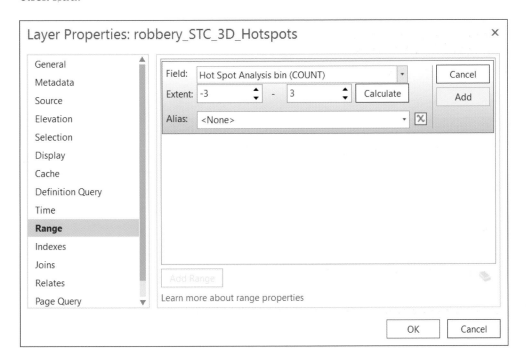

4. **Click OK.**

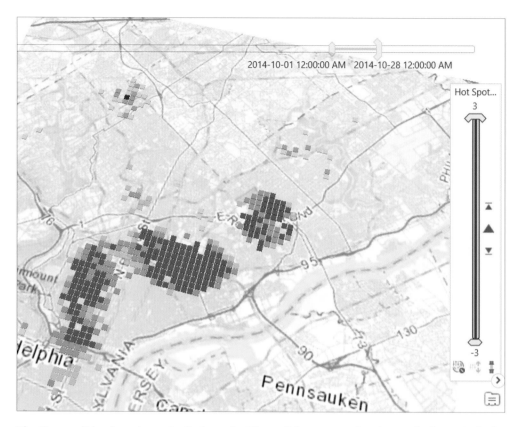

The Range slider functions similarly to the Time slider, except that instead of a period of time as the variable, it uses numerical values of any attribute that you are interested in. You will isolate only the highest confidence hot spots by using the values specified in the Emerging_Count_Hs_Bin attribute.

5. **On the Range tab, in the Full Extent group, enter a Min Value of** 2 **and a Max Value of** 3.

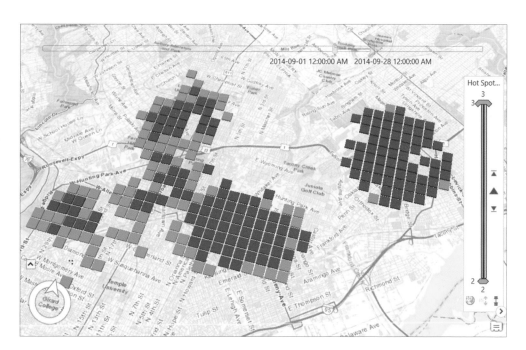

The scene shows hot spots that are now filtered by time *and* range. Using this combination of time and range, you can provide information for specific instances in time. Isolating hot spot clusters can provide a narrative as to when and where crimes happened.

TIP *The Range slider works in 2D and 3D, so you can use the same technique to easily filter the content you are viewing in a 2D map view as well.*

ON YOUR OWN

Optionally, explore connecting the Range slider to other numerical properties, such as Location ID.

6. **Save the project.**

Summary

You have learned what temporal data is, how to work with it, how to visualize it in 3D, and how to create an animation in ArcGIS Pro. The strength of GIS lies in the ability to recognize patterns and the organic behavior of crime—particularly in clusters over time. Although crime mapping is only one part of an often complex crime reduction plan, hot spot maps are useful in providing a spatial sense of the nature of high-crime areas and where crimes are concentrated. However, further analysis is required to understand why certain crimes happen more often in certain areas. The evidence of crime clusters can influence key decision-makers in crafting effective crime reduction tactics and can help crime analysts use GIS to convey that information.

Glossary terms

temporal data
operator
kernel density
hot spot analysis
space-time cube

CHAPTER 9
Determining suitability

Exercise objectives

9a: Prepare project data

- Convert a line feature to a polygon feature
- Clip raster layers
- Merge rasters

9b: Derive new surfaces

- Derive an aspect surface
- Derive a slope surface
- Derive a hillshade surface
- Visually compare analysis outputs

9c: Create a weighted suitability model

- Reclassify criteria rasters
- Combine criteria rasters

In this chapter, you will solve a common GIS problem: determining which areas are most suitable for a specific land-use purpose. To do so, you will take advantage of the raster data model.

As discussed in chapter 1, the first thing you should understand about a raster is that it is composed of a grid of *cells*, instead of discrete x,y coordinates, that define geographic entities. The cells contain values that are used to record and define geographic phenomena on the surface of the earth. Each raster cell represents a portion of the earth, such as a square meter or square mile, and usually has an attribute value such as elevation, soil type, or vegetation class associated with it. A surface, which is made up of raster cells, can be *discrete data*, which means that it shows distinct and discernible regions on a map, such as soil types, or it can be *continuous data*, which means that there are smooth transitions between variations in the range of data. Elevation data is one of the most common uses of a continuous raster. Typical raster datasets are extremely detailed and store a large amount of information, recording information about each cell.

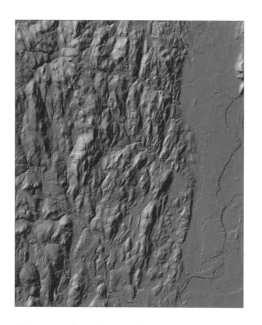

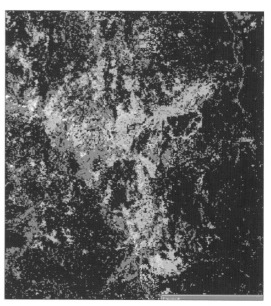

This raster of continuous data represents elevation. Data courtesy of the Office of Geographic Information (MassGIS), and the Commonwealth of Massachusetts Information Technology Division.

This raster of discrete data represents land cover. Data courtesy of Lake Tahoe Data Clearinghouse, US Geological Survey.

Map algebra—a language that combines GIS layers—is fundamental to raster analysis. You can evaluate raster cell values using functions and mathematical, logical, or Boolean operators. An output raster is the result of a cell-by-cell function performed on two input rasters.

A 3 × 3 grid added to another 3 × 3 grid produces a result grid, in which the cell values represent the total of the inputs.

Every raster cell has a value. Some cells may have a value of zero (for instance, representing no precipitation), and some cells may have a value of *NoData*, which means that no values were recorded for that cell. NoData cells are ignored in raster calculations.

The ArcGIS® Spatial Analyst extension contains close to 200 geoprocessing tools to perform comprehensive, raster-based spatial analysis. These tools allow you to perform basic mathematical and logical operations, as well as raster dataset creation and processing.

In this chapter, you will work with both vector and raster data, and you will use ArcGIS Pro standard tools as well as Spatial Analyst tools.

TIP *If you are using student trial software, the Spatial Analyst extension is automatically licensed.*

Scenario: suppose you are a vineyard manager. Your employer owns a large property in Monterey County, California. Much of it is already planted with grapes, but there are several unused blocks of land. You will use your enological knowledge and GIS skills to select the most suitable block for planting additional vines.

To grow wine grapes in a moderate coastal climate such as that of Monterey County, sunlight is the key ingredient—the more, the better, for rich fruit and high yield. Your specific criteria are that the land must be mostly south facing, for the longest sun exposure; the slope must be less than 14 percent, because you want the vineyard to be machine-harvestable; and finally, you want as little shade as possible. (For example, even if a block is south facing, it can have high slopes on either side, creating undesirable shade in the morning and afternoon.)

With your understanding of topography and GIS, you know that you must create and compare three layers: aspect, slope, and hillshade. You will have to do some preprocessing before you dive into the analysis. Once you have analyzed the property, you will overlay a feature class showing existing vineyards so that you can assess the available areas and decide which one is most suitable for planting.

Note: a special thanks to Tyler Scheid, Jonathan Vevoda, and Greg Gonzales at Scheid Vineyards for allowing us to use their data and general workflow. (Check out the fruit of their GIS-enabled efforts at www.scheidvineyards.com.)

DATA

In \\EsriPress\GTKAGPro\Vineyard\Vineyard.gdb:

- ned_1 and ned_2—elevation rasters (DEMs) for the study area *(Source: National Elevation Dataset from USGS)*
- planting_sites—a polygon feature class that outlines available planting areas (Source: Scheid Vineyards)
- property_boundary—a line feature class that delineates the property boundary of Scheid vineyards in Monterey *(Source: Scheid Vineyards)*
- Soils_clip—a soils survey layer that is clipped to the extent of the property_boundary feature class *(Source: USGS Soils Survey)*
- vineyard_blocks—a polygon feature class that shows where vineyards currently exist *(Source: Scheid Vineyards)*

To get the data in its current state, we downloaded elevation rasters (format: digital elevation model, or DEM) from https://viewer.nationalmap.gov/viewer. The property boundary coincides with two large DEM files, named ned19_n36x25_w121x00_ca_centralcoast_2010 and ned19_n36x25_w121x25_ca_centralcoast_2010. Once retrieved, the rasters' extents were then reduced to minimize student download size. (The extents were minimized by hand-digitizing a rough study area polygon around the property boundary, and then using the Extract by Mask tool to clip a portion of the full-sized rasters. The study area was used as a *mask*—a means of identifying areas to be included in a geoprocessing operation). You will perform a similar operation in exercise 9a.

EXERCISE 9A

Prepare project data

Estimated time to complete: 20 minutes

In this exercise, you will prepare data for a vineyard suitability analysis.

Exercise workflow

- *Convert the line property boundary feature to a polygon feature so that it can be used as an extraction mask.*
- *Using the Extract by Mask tool, reduce the extents of the two elevation rasters so that they are clipped to the extent of the property boundary.*
- *Append one extracted raster to the other, resulting in a single raster that is coincident with the property boundary.*
- *Project the elevation raster so that it is in the same coordinate system as the property boundary.*

Convert a line feature to a polygon feature

1. **In ArcGIS Pro, open Vineyard.aprx from your \\EsriPress\GTKAGPro\Vineyard folder.**

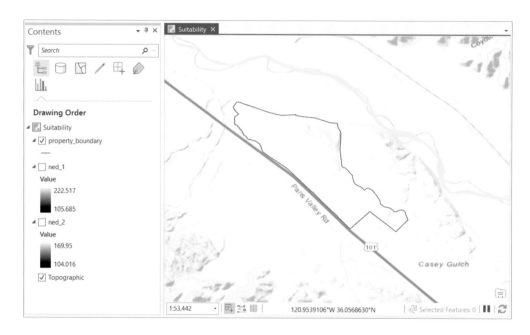

You see a basemap and a single line feature that demarcates the boundary of an approximately 1,100-acre property dedicated to viticulture.

2. **Turn on the ned_1 and ned_2 layers.**

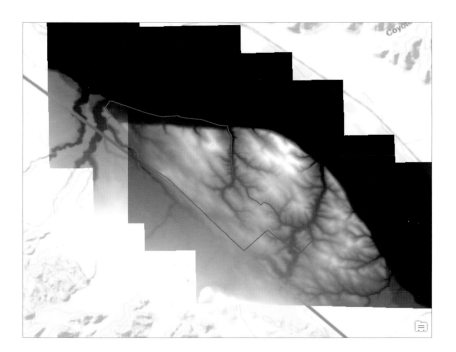

You will use the Extract by Mask tool to reduce the extent of the elevation layers to match the extent of the property boundary. However, to use the property boundary layer as a mask, it must be a polygon. Because it is currently a line feature (technically, three line features—you can tell by looking at the attribute table), you must first convert it to a polygon. For this task, you will use the Feature to Polygon geoprocessing tool.

3. **Open the Geoprocessing pane. Search for and open the Feature To Polygon tool.**

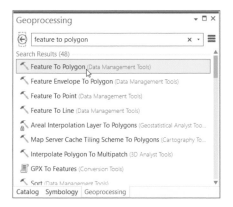

4. **For Input Features, click the prop-
   erty_boundary line feature class.**
   Notice that another input box appears—
   the tool provides the option to convert
   several feature classes at once, if needed.

5. **Point to the Output Feature Class
   box to see the default output
   location—the project geodatabase.
   Maintain this path, but change
   the Output Feature Class name to**
   property_boundary_poly.

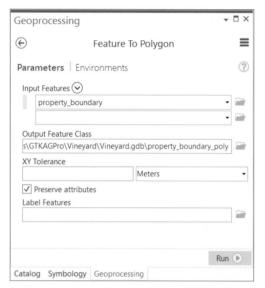

6. **Leave the optional parameters blank, and run the tool.** The new polygon layer is added
   to the map.

7. **Remove the line layer, and symbolize the new polygon layer however you like.**

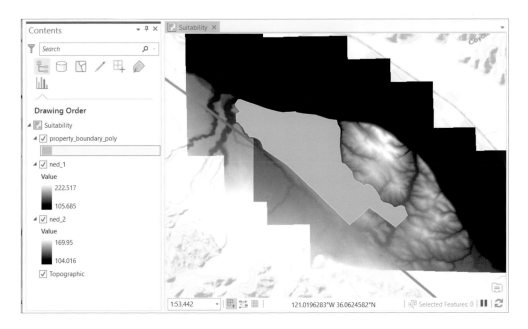

Clip raster layers

Now that the property boundary is a polygon, you can move forward with clipping the elevation rasters.

1. **In the Geoprocessing pane, search for and open the Extract by Mask tool.**

TIP *If the Geoprocessing pane still shows the Feature To Polygon tool, click the back arrow to return to the search function.*

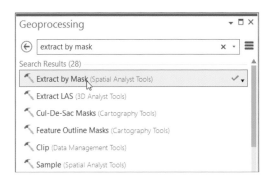

2. **For Input raster, click ned_1. For Input raster or feature mask data, click property_boundary_poly.**

3. **Maintain the Output raster name (notice that the output goes to the project's geodatabase—you can see the path when you hover over the output box).**

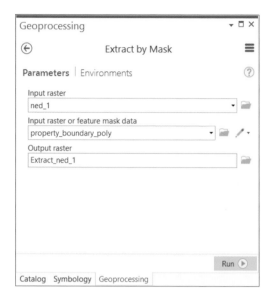

4. **Run the tool. When the tool is finished running, turn off property_boundary_poly and ned_1.**

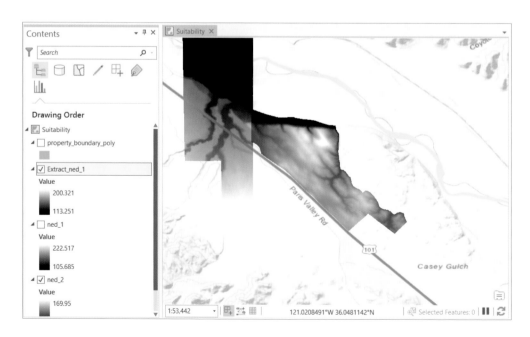

You have created a new raster whose shape matches the property boundary. Now you will perform the same operation on the second elevation raster.

> **TIP** *Another way to reduce the analysis extent is to use the Clip tool in the Data Management toolset. However, the Clip tool supports only an extent rectangle, not a polygon mask.*

5. **Rerun Extract by Mask using ned_2 as the input raster and keeping property_boundary_poly as the feature mask data. Rename the output raster to** Extract_ned_2.

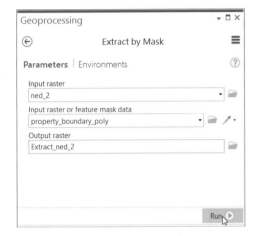

6. **Turn off ned_2, and zoom in slightly.**

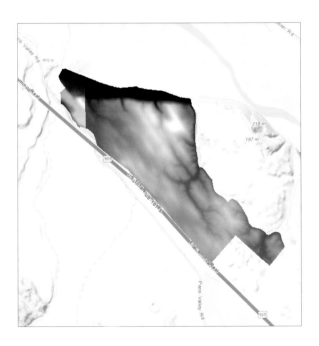

Merge rasters

Next, you will merge Extract_ned_1 and Extract_ned_2 into a new raster dataset, so that you can run analysis operations once, not twice.

1. **Find and open the Mosaic to New Raster tool.**

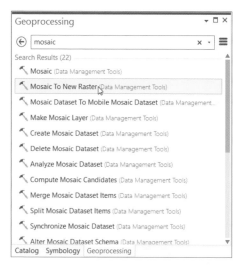

2. **In the tool, set the following parameters:**
    -   **For Input Rasters, choose Extract_ned_1 and Extract_ned_2.**
    -   **For Output Location, navigate to Vineyard.gdb.**
    -   **For Raster Dataset Name, enter** ned_property.
    -   **For Spatial Reference, choose property_boundary_poly.**
    -   **For Number of Bands, enter** 1.

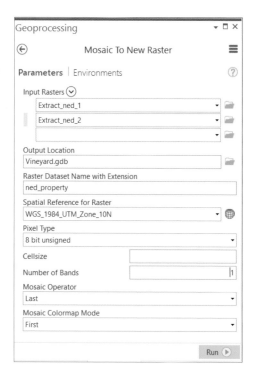

TIP *Why choose property_boundary_poly for the spatial reference? If you check the layer properties for the ned rasters, you will see they are in a geographic coordinate system, whereas if you check the layer properties of the property boundary, you will see it has both a geographic and projected coordinate system.*

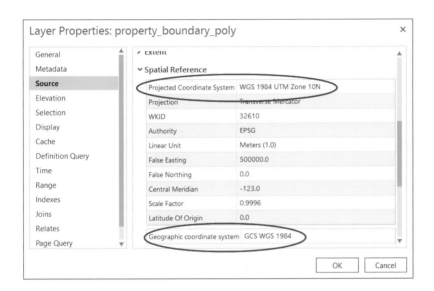

TIP *Before performing analysis, it is best practice to project your data to a local coordinate system. Recall that projections are discussed in chapter 4. If you want to convert any 2D data to 3D, you must also project to a local coordinate system. With the Mosaic To New Raster tool, you are combining two rasters into a new raster dataset, and you are projecting the output as well.*

TIP *Learn more about the optional parameters by pointing to the information icons to read the ToolTips.*

3. **Run the tool. When processing is complete, remove all elevation layers except ned_property.** Notice that this new layer is a combination of the two inputs.

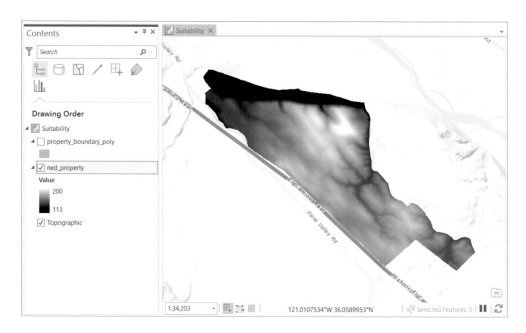

Your preprocessing is complete. What remains is one elevation raster, clipped to the extent of the study area and projected appropriately.

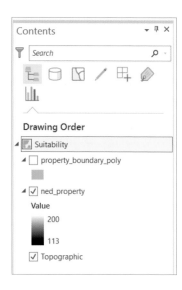

4. **Save the project.** You are finished with this exercise. The next exercise builds on the Vineyard project, so if you are continuing to exercise 9b, you can leave it open.

GIS IN THE WORLD: GIS AND PRECISION AGRICULTURE

Scheid Vineyards uses GIS not only for suitability analysis, but also incorporated GIS and GPS tracking technology into nearly every aspect of its operation. From tracking and managing worker productivity and equipment locations to detecting and mapping harmful pests such as insects and mold, its comprehensive geospatial technology program informs decision-making from the ground up.

Courtesy of Heidi Scheid.

Read more about Scheid Vineyards' innovative approach to viticulture here, "World-Class Vineyard Uses GIS to Fine-Tune All Its Operations": www.esri.com/esri-news/arcnews/fall13articles/world-class-vineyard-uses-gis-to-finetune-all-its-operations.

EXERCISE 9B

Derive new surfaces

Estimated time to complete: 30 minutes

Now you are ready to put some precision agriculture techniques to the test.

Exercise workflow

- *Create an aspect surface to determine which areas of the property are south facing (the preferred aspect for a vineyard).*
- *Create a slope surface to determine which areas of the property have slope measurements of 14 percent slope or less (required for machine harvesting).*
- *Create three hillshade surfaces using almanac values for sun altitude and azimuth to determine which areas are in sun with minimal shadows at three different times of the afternoon in mid-September (longer sun ripening just before harvest is ideal).*
- *Overlay polygon features that show which areas are already planted to reveal the areas that are available for a new vineyard. Assess the suitability of these areas based on your aspect, slope, and hillshade layers.*

Derive an aspect surface

1. **Continue working in the Vineyard project that you started in exercise 9a.** Assuming that you completed exercise 9a, your project has a map named Suitability, which looks like the graphic. The original two elevation rasters have been reduced to the extent of the property boundary (the project study area), merged into a single raster, and projected into a local coordinate system. The project includes a folder connection to the project folder, and the geoprocessing history reflects the work you have done thus far.

TIP *If you did not complete exercise 9a or are not sure if you did it correctly, you can use Vineyard_ B.aprx, found in the \Vineyard\Results\Vineyard_B folder.*

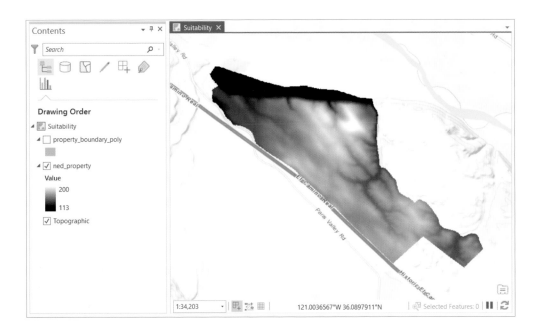

First, you will analyze the *aspect* of the topography (ground surface relief), to determine which direction each part of the ground primarily faces—north, south, east, west, or in between.

2. **In the Geoprocessing pane, find and open the Aspect tool (there are two— open the one from the Spatial Analyst toolbox).** The Aspect tool in the 3D Analyst toolbox is the same tool. There is quite a bit of overlap between the Spatial Analyst and 3D toolboxes. This workflow can be accomplished using either extension.

3. **For Input raster, click ned_property. For Output raster, maintain the default location, but edit the name to read** Aspect_ned. **Then run the tool.**

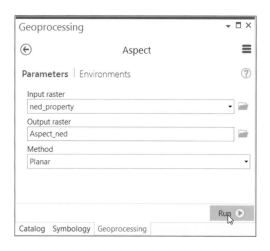

4. **When processing is complete, examine the results. Notice the areas that face south, southeast, or southwest.**

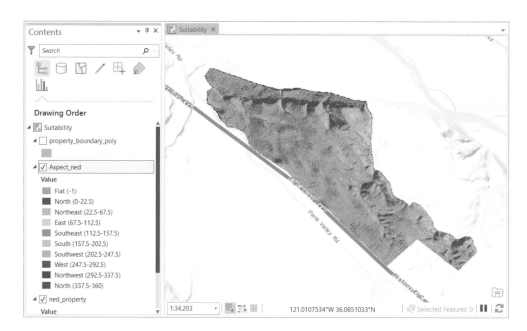

Next, you will derive the slope layer.

Derive a slope surface

1. **Find and open the Slope tool from the Spatial Analyst toolbox.**

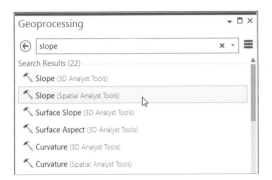

2. **In the tool, set the following parameters:**
   - **For Input raster, click ned_property.**
   - **For Output raster, maintain the default location, but edit the name to read** Slope_ned.
   - **Change the Output measurement to Percent rise.**
   - **Maintain the default Z factor,**

3. **Run the tool.**

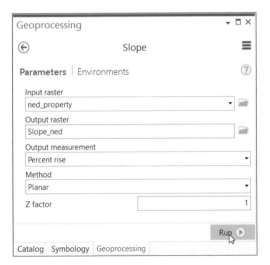

4. **When processing is complete, examine the results.** You are interested in seeing only whether the slope is less than or greater than 14 percent, so you will symbolize the layer to have only these two categories.

5. **With Slope_ned selected in the Contents pane, open the Symbology pane (Appearance > Rendering > Symbology button).**

6. **In the Symbology pane, change the number of classes to 2.**

   TIP *If you cannot change the number of classes, change the method to Equal Interval, the number of classes to 2, and the Method back to Manual Interval.*

7. **Click the color bar and choose Show Names, and then choose the Slope color scheme.**

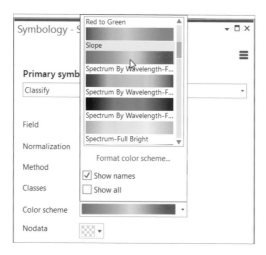

8. **At the bottom of the pane, under Class breaks, click the first Upper value box once to select it, and click again to make it editable. Replace the existing value with** 14, **and then press Enter.** Notice that when you change the class break values, the classification method changes from the default (Natural Breaks) to Manual Interval. You can see that although much of the property meets the criteria of having a slope that is less than or equal to 14 percent, a fair amount of it still has a greater percentage rise, which is not ideal for a new machine-harvestable vineyard.

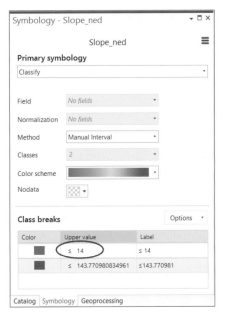

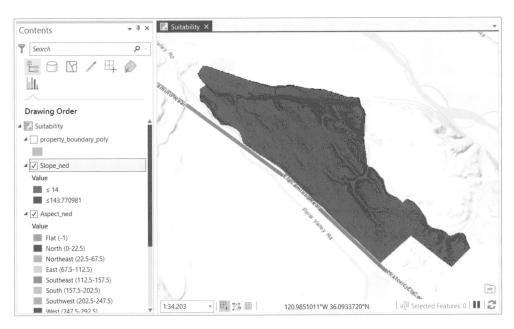

Derive a hillshade surface

A *hillshade* is a surface layer that depicts shadows to model the effect of an illumination source (usually the sun) over the terrain of the land. To create a hillshade, you must enter an azimuth and altitude value. *Azimuth* is the direction of the sun, expressed in positive degrees from 0 to 360, measured clockwise from north. The default azimuth is 315 degrees. *Altitude* is the angle of the sun above the horizon. The altitude is expressed in positive degrees, with 0 degrees at the horizon and 90 degrees directly overhead. In other words, relative to any given surface, azimuth tells you where in the sky the sun is, and altitude tells you how high. These values change depending on the time of day and time of year.

Your criteria indicate that you want lots of sun, especially during the critical fruit-ripening stage just before harvest. Harvest is usually sometime in October, depending on various factors. To be precise, you will look up almanac values for this area on September 15, at 2 p.m., 3 p.m., and 4:30 p.m., to show which areas have the longest sun exposure at this time.

1. **From the Spatial Analyst toolbox, find and open the Hillshade tool.**

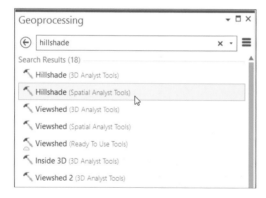

2. **For Input raster, click ned_property. For Output raster, maintain the default location, but edit the name to read** Hillshade_ned_1. You will obtain the azimuth and altitude values from the US Naval Observatory Astronomical Applications Department. You will search for values for King City, California, the nearest major town.

3. **Open a web browser, and go to http://aa.usno.navy.mil/data/docs/AltAz.php.**
   (Alternatively, search for usno azimuth table.**) In Form A, set the following options:**
   - **Object: Sun**
   - **Year/Month/Day:** 2015; **September;** 15
   - **Tabular interval:** 10 **minutes**
   - **State or Territory: California**
   - **City or Town Name:** King City

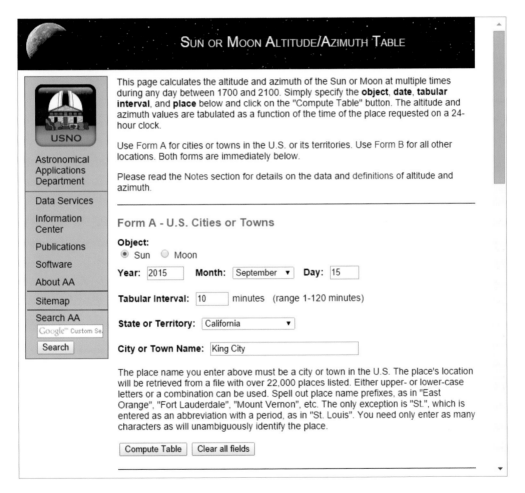

4. **Click the Compute Table button. Take a look at the table. Notice that the hours are listed in military time, from 6:00 a.m. (06:00) to 8:00 p.m. (20:00).** The second column shows altitude, and the third column shows azimuth.

```
Astronomical Applications Dept.
U.S. Naval Observatory
Washington, DC 20392-5420

KING CITY, CALIFORNIA
    o   ,    o   ,
W121 07, N36 13

Altitude and Azimuth of the Sun
Sep 15, 2015
Pacific Daylight Time

                Altitude      Azimuth
                              (E of N)

  h  m            o              o
06:00           -10.2          78.6
06:10            -8.2          80.1
06:20            -6.2          81.7
06:30            -4.2          83.2
06:40            -2.2          84.6
06:50             0.3          86.1
07:00             2.1          87.6
07:10             4.0          89.1
07:20             6.0          90.6
07:30             8.0          92.0
07:40            10.0          93.5
07:50            12.0          95.1
08:00            14.0          96.6
08:10            16.0          98.1
08:20            17.9          99.7
08:30            19.9         101.3
```

5. **In the table, look up the values for 2:00 p.m. (14:00).**

```
12:00       54.0        154.1
12:10       54.8        158.1
12:20       55.5        162.3
12:30       56.0        166.7
12:40       56.4        171.1
12:50       56.6        175.6
13:00       56.7        180.1
13:10       56.6        184.7
13:20       56.4        189.2
13:30       56.0        193.6
13:40       55.4        197.9
13:50       54.7        202.1
14:00       53.9        206.2
14:10       52.9        210.0
14:20       51.9        213.7
14:30       50.7        217.3
14:40       49.4        220.6
14:50       48.1        223.8
15:00       46.6        226.8
15:10       45.1        229.7
15:20       43.6        232.4
```

TIP *You may want to write the values on a piece of paper to ensure accuracy.*

6. **Enter these values in the Hillshade tool. (Remember that the table lists altitude first, which is the second value in the tool.) Maintain the remaining default settings, and run the tool.** The output is a shaded relief layer that shows which areas are illuminated and which are in shadow at 2 p.m. in mid-September. A raster cell value of 0 represents complete shadow, with no illumination from the sun. A raster cell value of 255 represents full illumination.

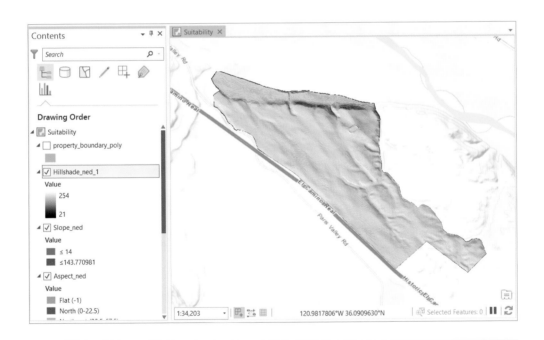

Is any part of the property in complete shadow at this time?

7. **Repeat the hillshade process two additional times, using the almanac values for 3:00 p.m. (15:00) and 4:30 p.m. (16:30). Name the output rasters** Hillshade_ned_2 **and** Hillshade_ned_3, **respectively.**

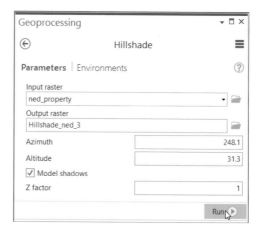

The illuminated areas decrease as the sun descends in the sky.

Hillshade rasters are natural candidates for 3D display.

8. On the **View** tab, click the **Convert** button, and choose to convert to a local scene.

9. When the **Suitability_3D map is created, use the Explore tool to rotate and tilt the map so you can see the relief.**

REMIND ME HOW

Press and hold the scroll wheel on the mouse to rotate and tilt a 3D map.

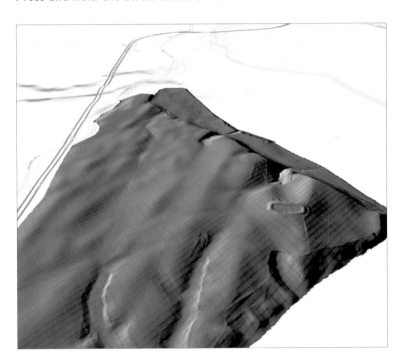

You will complete the chapter in the 2D map, but feel free to examine the 3D map at any time.

Visually compare analysis outputs

1. **In the Suitability map (the 2D map), add vineyard_blocks from Vineyard.gdb. Symbolize it as you wish.** This layer represents already planted vineyard blocks. Next, you would typically hand-digitize a layer of potential planting sites using ArcGIS Pro editing tools and your knowledge of the property, but to save time we have provided the layer for you.

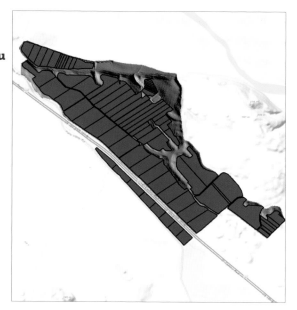

2. **Add planting_sites from Vineyard.gdb, and symbolize it using a black 1-point outline and hollow fill for easy visibility.** Indicated with red arrows in the graphic, these unplanted areas of the property are available for future plantings. First, you will assess the slope criteria because machine harvesting (rather than picking by hand) is a requirement in this scenario.

3. Turn off the three hill-
    shade layers, and look
    for future planting sites
    that have mostly low-
    slope land (lots of green
    shading).

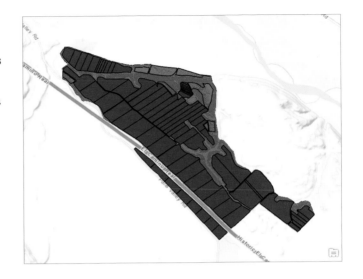

How many planting sites contain mostly low-slope
(less than 14 percent) topology?

4. Turn off Slope_ned to
    reveal the Aspect_ned
    layer.

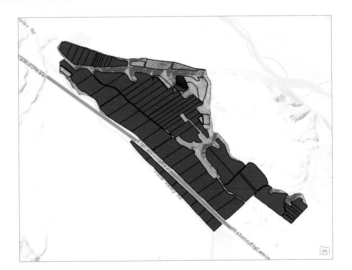

How many potential planting sites include at least some
land that faces south, southeast, or southwest?
(Zoom in and pan as needed.)

5. **Turn on hillshade_ned_1.**

Are any of the potential planting sites in shadow at 2:00 p.m.
in mid-September?

---

6. **Turn on hillshade_ned_2. Notice that the shadows intensify over most of the planting sites.**

7. **Turn on hillshade_ned_3. Notice that the shadows increase further in most areas.**

Can you identify a planting site that meets the slope and aspect
criteria and also has decent sun exposure at 4:30 p.m., thus
revealing the best site to plant the new vineyard?

Rarely will a site fit all of your suitability criteria perfectly. In exercise 9c, you will combine your criteria in a weighted model for a more precise analysis.

8. **Save the project.** If you are continuing to exercise 9c, keep the project open. Otherwise, it is okay to close ArcGIS Pro and come back to it later.

Create a weighted suitability model

Estimated time to complete: 20 minutes

You will find the most suitable planting site by combining the values of three layers (aspect, slope, and the third hillshade that shows sun illumination in the latest part of the afternoon). Before the layers are combined, they must have a common range. Combining them as they are now would produce nonsensical results. Consider: How does a slope of 9 compare with a hillshade value of 190? Instead, you will *reclassify* the values so that you can add them together. (That means you will replace raster cell values with new cell values so that the rasters can be logically combined.) Each layer will be weighted—that is, as a percentage of the whole. In this way, you can prioritize your criteria. For example, if having the correct slope is most important, you will give slope the highest weight.

Exercise workflow

- *Create a new geoprocessing model. Add three Reclassify processes, in which you will reclassify three criteria surfaces: Aspect_ned, Slope_ned, and Hillshade_ned_3. Replace the original cell values with either 1 (for desirable values) or 0 (for undesirable values).*
- *Add a Raster Calculator process to the model. Add together the reclassified values of the three inputs, but multiply each input by a percentage of 100. In this way, your end results will be weighted according to criteria priority.*

Reclassify criteria rasters

1. **In ArcGIS Pro, continue working with Vineyard.aprx.** Assuming that you completed the previous exercises in this chapter, your Suitability map contains the topographic basemap and nine layers, including three feature classes (planting_sites, vineyard_blocks, and property_boundary_poly, which is turned off), plus three hillshade surfaces, a slope surface, an aspect surface, and the ned_property elevation surface from which the other surfaces were derived.

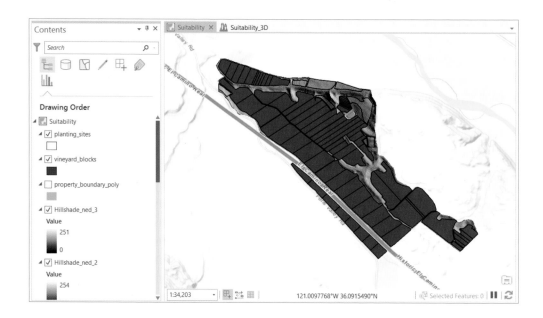

> **TIP** *If you did not complete exercises 9a and 9b or are not sure if you did them correctly, you can use Vineyard_ C.aprx, found in the \Vineyard\Results\ Vineyard_C folder.*

2. **On the Analysis tab, click the ModelBuilder button to create a new geoprocessing model in the project toolbox.** Before you can combine the criteria into a single output, they must be reclassified so that their cell values make sense in a calculation. In this case, you will change the desirable cell values to 1, and the undesirable values to 0.

3. **Search for the Reclassify tool (Spatial Analyst Tools) in the Geoprocessing pane, but do not open it. Drag the tool from the list of search results into the model three times.**

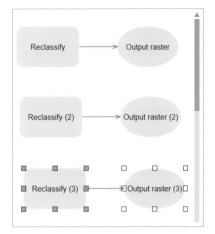

TIP *Notice that the ModelBuilder tab has a Tools button, which allows you to open the Geoprocessing pane from there.*

First, you will reclassify the aspect surface.

4. **Take a look at the Aspect_ned legend.** The desirable values are Southeast (112.5–157.5), South (157.5–202.5), and Southwest (202.5–247.5). You will also consider flat surfaces (-1) as desirable.

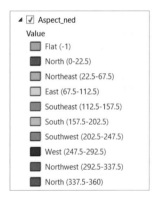

5. **In the model, double-click the first Reclassify tool to open it.**
   - **For Input raster, click Aspect_ned.**
   - **Maintain the Reclass field of Value.**
   - **Give the desirable traits a new value of** 1, **and the others a new value of** 0. **(See the graphic for guidance.)**

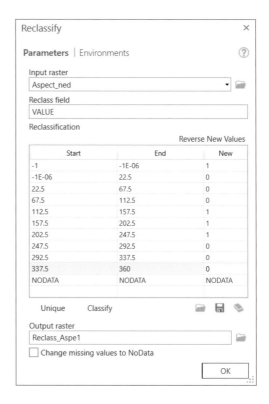

Reclassify			

Parameters | Environments

Input raster
Aspect_ned

Reclass field
VALUE

Reclassification

Reverse New Values

Start	End	New
-1	-1E-06	1
-1E-06	22.5	0
22.5	67.5	0
67.5	112.5	0
112.5	157.5	1
157.5	202.5	1
202.5	247.5	1
247.5	292.5	0
292.5	337.5	0
337.5	360	0
NODATA	NODATA	NODATA

Unique   Classify

Output raster
Reclass_Aspe1

☐ Change missing values to NoData

OK

TIP *Be careful that your screen matches the graphic exactly, or your final output may be incorrect.*

6. **Maintain the default output, and click OK. The tool is now ready to run in ModelBuilder.**

Aspect_ned → Reclassify → Reclass_Aspe1

TIP *Resize or move the model elements as you prefer. You can also click the Auto Layout button on the ModelBuilder tab under View to arrange model elements automatically.*

7. **Open the second Reclassify tool in the model.**
   - **For Input raster, click Slope_ned.**
   - **Maintain the other parameters, but for the 14-143.770981 category, change the New value to** 0.

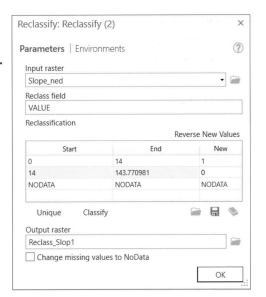

Reclassify: Reclassify (2)

Parameters | Environments

Input raster
Slope_ned

Reclass field
VALUE

Reclassification

Reverse New Values

Start	End	New
0	14	1
14	143.770981	0
NODATA	NODATA	NODATA

Unique   Classify

Output raster
Reclass_Slop1

☐ Change missing values to NoData

OK

8. **Click OK to make the tool ready in the model.** The second tool is ready to run.

9. **Open the third Reclassify tool. For Input raster, click Hillshade_ned_3.** Notice that the value table currently has no values. The reason is because the hillshade layers are symbolized using a stretched symbology scheme, so there is not a category for each range of values. You know that some shade is unavoidable at 4:30 in the afternoon, yet you want to rule out areas that have little to no sunlight. Knowing that 0 means no illumination and 251 is the highest level of illumination at this time (look at the legend), you decide to draw the "unacceptable" cutoff value at 90.

10. **In the first Reclassification cell under Start, enter** 0. **Under End, enter** 90; **and under New, enter** 0.

11. **Press Enter to start a new row. In the next row, under Start and End, enter values of** 91 **and** 251, **respectively. Under New, enter** 1.

12. **Make sure that your screen matches the graphic, and then click OK.** All of the Reclassify tools are ready to run.

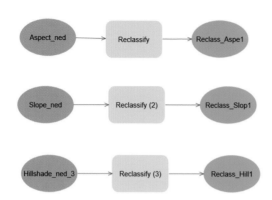

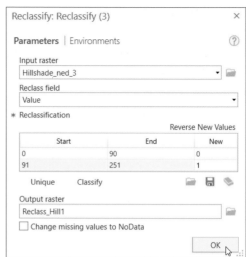

Combine criteria rasters

1. **Search for the Raster Calculator tool in the Geoprocessing pane (Spatial Analyst toolset) and add it to the model.**

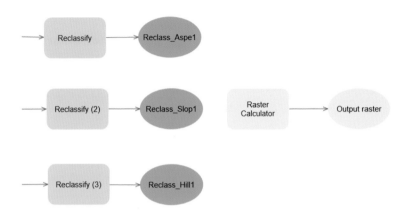

You decide to weight aspect, slope, and hillshade to 20 percent, 50 percent, and 30 percent, respectively.

2. **Open the Raster Calculator tool. Double-click Reclass_Aspe1 to add it to the expression box.** In the expression box, it looks like "%Reclass_Aspe1%". This output raster is from the operation that will reclassify the aspect layer.

3. **Place your cursor after the final quotation mark, and then double-click the multiplication operator (*) to add it to the equation. Enter a space, and then enter** .20.

4. **Double-click the addition operator** (+)**.** So far, the expression reads:
"%Reclass_Aspe1%" * .20 +

5. **Double-click Reclass_Slop1 to add it to the expression. Double-click the multiplication operator, enter a space, and then enter** .50.

6. Double-click the addition operator.

```
"%Reclass_Aspe1%" * .20 + "%Reclass_Slop1%"
* .50 +
```

7. Double-click Reclass_Hill1. Double-click the multiplication operator, enter a space, and then enter .30. Click OK.

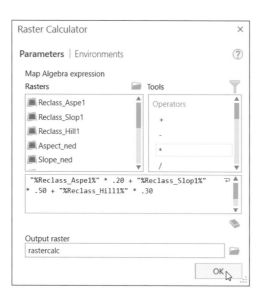

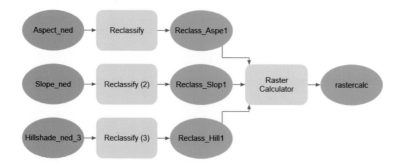

8. **Save and then run the model.**

9. **When processing is complete, close the progress window.**

10. **In the Catalog pane, notice that a new raster named rastercalc is in your project geodatabase. (You may need to refresh the geodatabase first.)**

11. **Click the Suitability tab to see the map.**

12. **From the Catalog pane, drag rastercalc to the map below the planting_sites layer. Turn off vineyard_blocks.**

13. **Symbolize rastercalc using classified symbols, Natural Breaks method, 3 classes.**

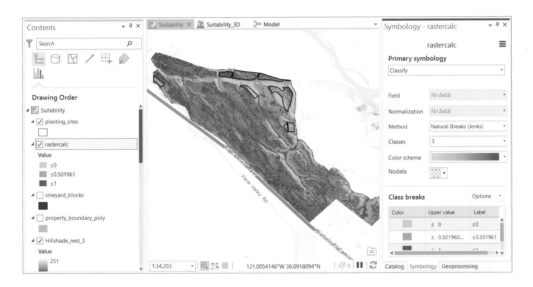

There are three top planting
candidates.

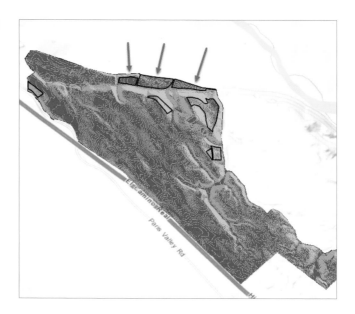

14. Save the project.

The GIS analysis is an important first step, but it is just one piece of the equation. You know
that these sites are geographically suitable for planting, but now the owners must consider
several other factors before investing in planting. For instance:

- What grape varietal makes the most sense from a business perspective? (Syrah,
  pinot noir, and chardonnay are common varietals in this region.)
- What is the average tonnage yield per acre? (Because of your scrupulous suitability
  analysis, the yields are likely to be at least average, likely better.)
- Are these grapes going to be sold to other winemakers or used for the owner's
  label?
  - If the former, what is the average price per ton that they can expect to reap?
  - If the latter, how many cases is the owner's winery able to get per ton of grapes,
    and what is the profit margin?
- How much will the additional vineyard blocks add to the overhead costs of field-
  workers, vineyard managers, fertilizer, water, pest abatement, and so on?
- Will the financial gain in good years be enough to offset losses from the inevitable
  challenges of drought, pests, or early frost?
- What type of soil exists at each site, and how does that type affect the decision?
  (Soils are important in terms of rootstock selection, which down the line will dic-
  tate your irrigation practices.)

> ## ON YOUR OWN
>
> *Add Soils_clip from Vineyard.gdb. Symbolize it using unique values based on the Name value field. Place the planting_sites layer at the top of the Contents pane.*

What soil type(s) are represented in your top-choice planting sites?

The work you have done thus far, using ArcGIS Pro and Spatial Analyst, provides an excellent starting point in the decision-making process with the vineyard management team.

> ## ON YOUR OWN
>
> *Switch out the topological basemap for an imagery basemap, and turn off all layers except planting_sites.*

Have any of the potential sites been planted?

You are finished with this chapter. If you are not continuing to chapter 10, it is okay to close ArcGIS Pro.

Summary

This chapter gave you an introduction to precision farming using GIS. Using raster elevation data and the Spatial Analyst extension, you created aspect, slope, and hillshade layers. Then you learned how to create a symbol model that combined the three criteria to determine the best planting site. Remember that before combining rasters using the Raster Calculator tool, you will likely have to reclassify them to give them a common value scheme. You can see that for some GIS purposes, Spatial Analyst is an indispensable tool.

Glossary terms

cell

discrete data

continuous data

map algebra

NoData

mask

aspect

hillshade

azimuth

altitude

reclassify

## CHAPTER 10

Presenting your project

Exercise objectives

10a: Apply detailed symbology

- Create a definition query
- Fine-tune the symbology
- Apply symbol layer drawing

10b: Label features

- Label features using the Maplex Label Engine

10c: Create a page layout

- Add a layout to the project
- Insert map frames
- Insert a legend element
- Insert a scale bar
- Insert a north arrow
- Insert dynamic text, a title, and rectangles

10d: Share your project

- Export a layout to a PDF
- Save a layout file
- Package and share a project template

Maps convey your ideas about locations into the minds of map readers. It is a fine balance of communicating the appropriate amount of information by displaying data and graphic elements through a suitable medium, such as a paper map or mobile device.

Every *layout* has a collection of organized elements. In a layout, you can arrange common elements, including the map, map labels, a title, legend, scale bar, north arrow, captions, and additional graphics. Creating a page layout requires answering some key questions:

- What is the purpose and intended use of the map?
- Who is the target audience, and what is their familiarity with the area of interest? What kinds of labels will provide a sense of familiarity?
- What is the most appropriate map scale to convey the map information? How detailed or general will the map data be displayed?
- What type of map is it—quantitative or qualitative? What kind of symbology will most effectively communicate the data?
- What are the dimensions of the map (printed or digital)? What display orientation will best fit, and how will the map elements be organized?

A good map not only provides facts, but also persuades opinion. You will want to think about the context of the data and frame the results to match the geographic area of the phenomena. Also think about graphic hierarchy and how map features are overlaid on top of each other. A well-composed layout promotes a visual hierarchy that map readers can explore naturally. Remember that print maps are just one way to share work and that ArcGIS Pro provides other options, including sharing as a PDF, a web map, or a web scene in the case of 3D.

In this chapter, you will compose a layout that consists of several maps of broadband internet availability for two counties in Utah.

Scenario: the agency you work for aggregates much of the broadband accessibility data for the state of Utah. As part of the agency's ongoing activities, it will determine broadband internet availability and work with stakeholders and service providers to identify accessibility, support broadband planning and policy, and study broadband trends.

Much of the broadband internet availability data has been collected through broadband service provider submissions and publicly available sources. As part of a large mapping project, several areas must be mapped. You will use your knowledge of cartographic principles and GIS skills to complete maps that show the maximum advertised broadband speeds for a county in Utah. In addition, you will prepare some data before symbolizing and labeling it. Once you have identified which areas to map, you will create several layouts to complete the mapping project, which can then be exported and shared online.

GIS IN THE WORLD: TELECOMMUNICATIONS

TK Telekom is a dynamically developing telecommunications operator that uses Esri and ArcGIS solutions to efficiently manage its network of nearly 30,000 kilometers of lines. The company provides internet services, telephone services, data transmission, and line lease for telecommunications operators, public administrators, and business customers. GIS operators are able to see the inventory system quickly to identify potential failure locations and enable implementation of precise remedies against a background of digital vector and raster maps. Read more about the project here, "Facilitating and Improving Telecom Network Management": www.esri.com/esri-news/arcnews/winter1213articles/facilitating-and-improving-telecom-network-management.

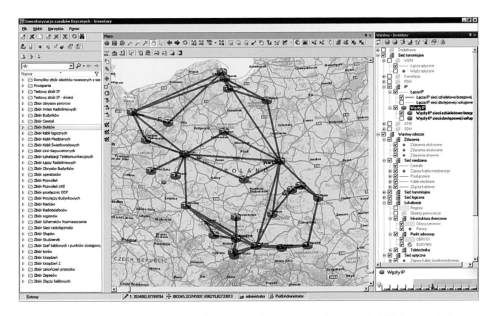

The SunVizion Network Inventory, based on ArcGIS for Server and Microsoft SQL Server platforms, visualizes this IP network. Courtesy of Suntech S.A.

DATA

In the \\EsriPress\GTKAGPro\Broadband\Broadband.gdb:

- Wireline Broadband—a polygon feature class of wired internet coverage *(Source: Utah Broadband Project; Utah Automated Geographic Reference Center; Governor's Office of Economic Development)*
- Wireless Broadband—a polygon feature class of wireless internet coverage *(Source: Utah Broadband Project; Utah Automated Geographic Reference Center; Governor's Office of Economic Development)*
- Community Anchor Institutes—a point feature class of the community sites *(Source: Utah Broadband Project; Utah Automated Geographic Reference Center; Governor's Office of Economic Development)*

EXERCISE 10A

Apply detailed symbology

Estimated time to complete: 20 minutes

In this exercise, you will apply detailed symbology to the data to visualize maximum advertised broadband speeds—both wired and wireless broadband. To view symbology and features, it is best to maximize your ArcGIS Pro window if possible.

Exercise workflow

- *Create a definition query to show only fixed wireless technology.*
- *Change symbology to symbolize only certain broadband download speeds.*
- *Apply symbol layer drawing to control the drawing order of wired broadband symbology.*

Create a definition query

1. **Open the Broadband.aprx project from the \\EsriPress\GTKAGPro\Broadband folder.** The project has two maps, the active one named Max Advertised Broadband Speed, and another map named Community Anchor Institutes, which you will use later. In the active map, the Wireline_Broadband layer represents all census tracts that have wired broadband technologies, and the Wireless_Broadband layer represents areas that have wireless technologies. You are required to show only fixed wireless technology and visualize it. To show only wireless, you will create a definition query to select a subset of all of the wireless technologies.

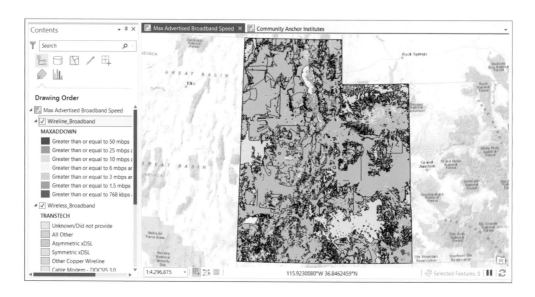

2. **Open the Symbology pane to view how the Wireless_Broadband layer was symbolized.** The Wireless_Broadband layer was symbolized using Unique Values and a random color scheme. This particular layer has its values stored as numbers in the geodatabase and wireless broadband technology label names associated with them. You are interested in Terrestrial Fixed Wireless technologies that have corresponding values of 70 and 71, Unlicensed and Licensed, respectively.

ON YOUR OWN

Open the Wireless_Broadband attributes, and explore the geodatabase fields, domains, and subtypes in the Fields View to get an idea of how the data is organized.

3. **With the Wireless_Broadband layer selected in the Contents pane, go to the Data tab on the ribbon, and click the Definition Query dialog box launcher.**

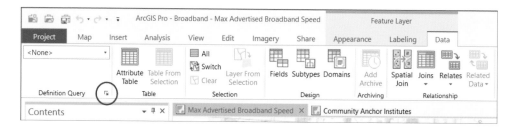

Creating a definition query expression to select a subset of features for the layer to display is slightly different from using the geoprocessing tool Select Layer By Attribute. Fundamentally, creating the expression is the same, but the latter results in selected features. In this workflow, only the features that satisfy the expression are displayed on the map but are not selected.

4. **In the Definition Query frame, click Add Clause, and create the following expression: Technology_of_Transmission is Equal to 70 – Terrestrial Fixed Wireless – Unlicensed. Click Add.**

5. **Add a second clause using an OR operator (or) that queries the value 71 – Terrestrial Fixed Wireless – Licensed.**

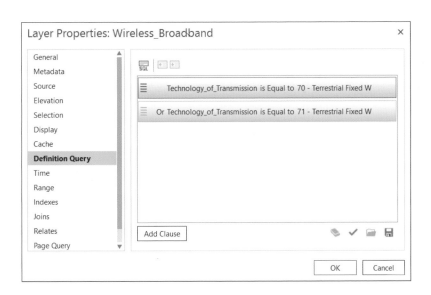

> **TIP** *Using the AND operator (and) will not work here, because in this dataset there are two distinct fixed wireless values in the Wireless_Broadband layer. When the OR operator is used, you can select from all points—that is, Terrestrial Fixed Wireless – Unlicensed or Terrestrial Fixed Wireless – Licensed. To view the SQL statement in full, click Switch to Edit SQL Mode.*

TRANSTECH is the field name of the field alias for Technology of Transmission.

6. **Click the "Verify the SQL expression is valid" button . (If the SQL expression was not verified successfully, retrace your steps, and try again.) Click OK.** A successful definition query results in only fixed wireless technology being displayed on the map.

7. **Turn off the Wireline_Broadband layer.**

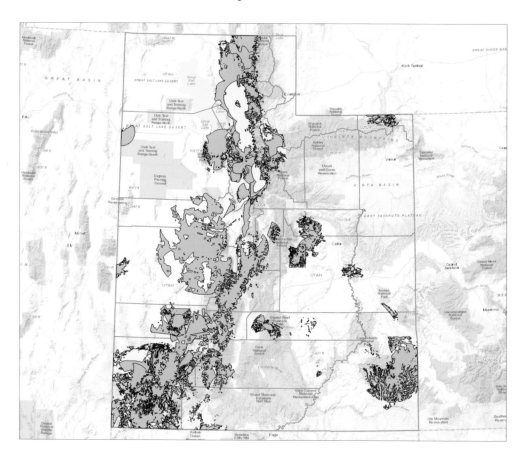

With the Wireline_Broadband layer turned off, you can see more clearly the Wireless_Broadband layer.

How many areas are of fixed wireless technology?

Fine-tune the symbology

After the definition query isolates only fixed wireless technology areas, apply more detailed symbology to the wireless broadband layer to show the maximum advertised download speed available. You will symbolize only broadband download speeds that are faster than 768 kilobits per second (Kbps).

1. **With the Wireless_Broadband layer selected in the Contents Pane, go to the Symbology pane and make sure Unique Values is chosen.**

2. **In Field1, select Maximum_Advertised_Downstream_Speed. (If necessary, adjust the Label column width to make the labels visible.)**

3. **Right-click the Value column heading, and click Sort Ascending.**

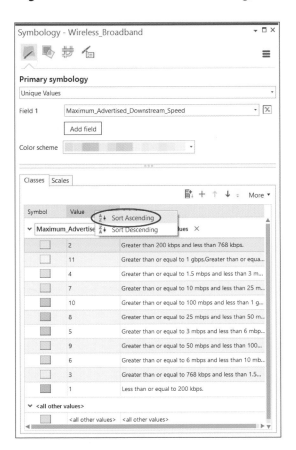

Eleven values are added. However, several values do not meet the requirement of symbolizing speeds of only 768 Kbps or faster. You will remove them and effectively hide the others.

4. **Right-click and remove value 1 (Less than or equal to 200 kbps).**

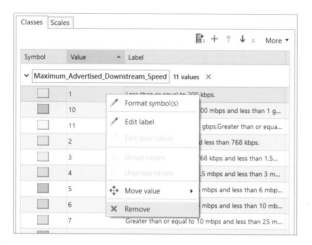

5. **Remove value 2 (Greater than 200 kbps and less than 768 kbps).** Next, organize the categories into a more logical order.

6. **Sort the Value column in descending order.**

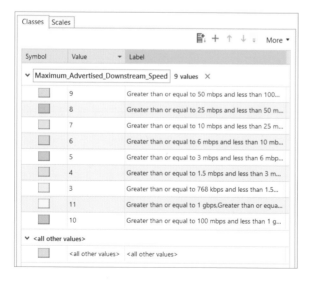

The values 10 and 11 are at the bottom of the list. Because they are both categorized as faster (gigabits per second [Gbps]) download speeds, you will move them to the top of the list.

7. **Select both values 10 and 11 (press and hold the Ctrl key, and click each row), and right-click. Then click Move value > Top.**

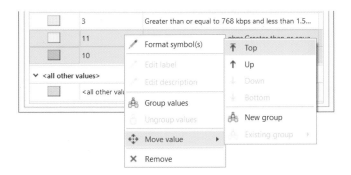

Nine values are a lot to symbolize. Next, you will group the three highest values together to reduce the number of total values to just seven values.

TIP *Remember that panes can be either docked or floating. You can adjust their width or height to make pane elements visible and easier to work with.*

8. **Select values 9, 10, and 11, and right-click. Then click Group values. Rename the group to** Greater than 50 mbps. **Rename the remaining values, using the following ranges in the text descriptions:**

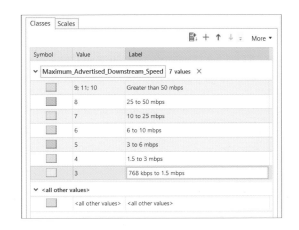

- **Value 8:** 25 to 50 mbps
- **Value 7:** 10 to 25 mbps
- **Value 6:** 6 to 10 mbps
- **Value 5:** 3 to 6 mbps
- **Value 4:** 1.5 to 3 mbps
- **Value 3:** 768 kbps to 1.5 mbps

The seven values have been prepared and are now ready for a different color scheme.

9. From the Color Scheme drop-down list, select the Show all check box. Scroll down, and select Red-Yellow-Green (7 classes).

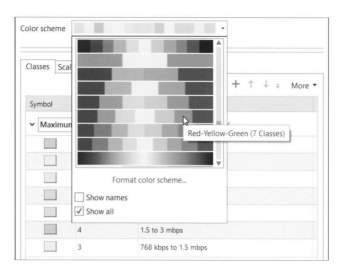

10. Click the More drop-down, and click Format all symbols.

11. Go to Properties, change the Outline color to No color, and click Apply.

12. Click the back button to return to the Wireless_Broadband Symbology pane. From the More drop-down list, select "Show all other values" to disable it.

13. Go to the Appearance tab, locate the Effects group, and adjust the Layer Transparency slider to 40%. (Alternatively, enter a value of 40, and press Enter.)

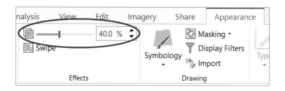

14. **Turn on the Wireline_Broadband layer.** The symbology for the Wireless_Broadband layer is complete. You should see the areas with wireline broadband more clearly now as well as the difference between wireline and wireless broadband (transparent in color).

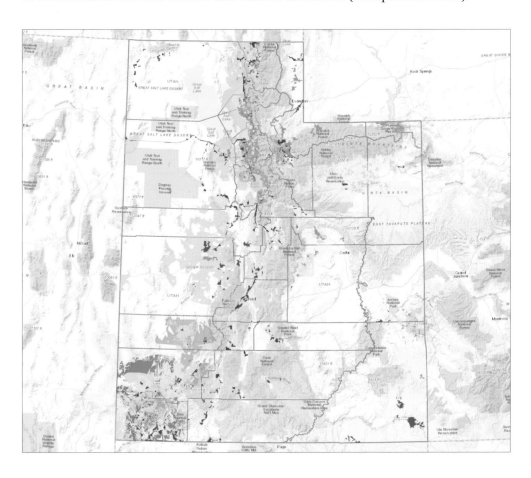

Apply symbol layer drawing

You will notice on the Wireline_Broadband layer that some of the higher speeds are not being displayed in the correct order (the slower speeds in green should be drawn below the higher speeds in red—true to the symbology color ramp). Next, you will look at the symbology on the map and compare it to the order of the color ramp in the Contents pane. It may be a bit difficult to see how overlapping features are being represented, but upon closer inspection (as shown in the graphic), the higher speeds are being drawn in a lower order.

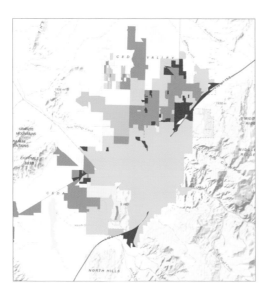 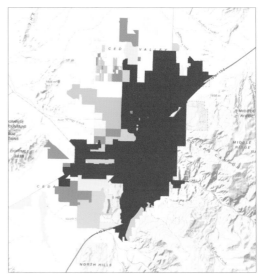

Left, no symbol layer drawing is used. *Right*, symbol layer drawing is enabled. The symbols are drawn according to the order specified in the Symbology pane.

You will use symbol layer drawing to control the drawing order of feature symbology—essentially overriding the default drawing sequence. This override is particularly helpful in this situation in which overlapping features exist.

1. **With the Wireline_Broadband layer selected in the Contents pane, go to the Symbology pane, and go to the Symbol layer drawing tab.**

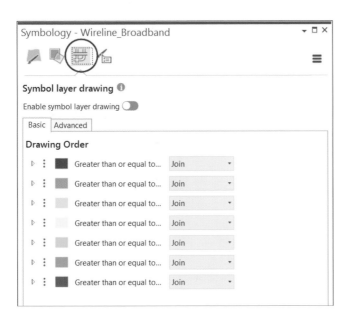

2. **Click the "Enable symbol layer drawing" toggle.**

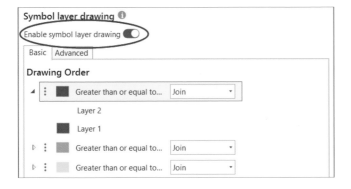

3. **Expand a group to see its contents.** You achieved the correct drawing order based on the order that was used in the Symbology pane, so no reordering is needed. If you need to reorder symbol layers, simply drag and drop the layers in the order desired.

4. Using steps 1 to 3, enable symbol layer drawing in the Wireless_Broadband layer.

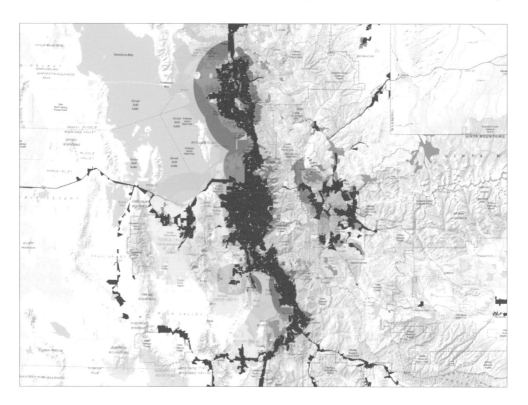

This map now shows coverage by fixed broadband technologies (DSL, cable, fiber, and fixed wireless) based on maximum advertised download speeds marketed to all consumer levels, including commercial business. Next, you will label features in another map before you use it in a page layout.

EXERCISE 10B
Label features

Estimated time to complete: 15 minutes

Your mapping project also requires features to be labeled on the Community Anchor Institutes map. *Labels* are based on one or more feature attributes and placed near or on a feature. As you may have seen in earlier chapters, ArcGIS Pro places labels for all features in a layer with a single click based on predetermined labeling rules. This process, known as *dynamic labeling*, is quick but has its drawbacks. Label positions can change depending on map scale, addition or removal of features, or the attribute used. Dynamic labels are still useful for most mapping projects. However, keep in mind that static layouts for print require extra attention so that labels are exactly where you want them.

The map has several layers, including a point layer that represents community libraries. It is symbolized based on whether there is public Wi-Fi available at each location. However, some libraries are symbolized as unknown. You will create labels so that your department can get a visual perspective of which libraries still require information to be collected so that the database can be updated accordingly.

MAPLEX LABEL ENGINE VERSUS STANDARD LABEL ENGINE

Maplex Label Engine is the default label engine, because it provides the best positioning and has finer control over labels, such as enhanced position strategy, conflict resolution, weighting, and priority. The Standard Label Engine is used to label as many features as it can without overlap and for cases in which the utmost speed is needed. To learn more about the differences between the two label engines, go to ArcGIS Pro Help > Maps > Author maps > Text > Add text to a map.

10

Exercise workflow

- *Use Maplex label settings to label libraries that have an unknown public Wi-Fi status.*

Label features using the Maplex Label Engine

When you use the Maplex Label Engine, you have access to a new set of label placement properties that allow you to control how labels are oriented, formatted, and placed in feature-dense areas, and how conflicts between labels can be resolved. In addition to the standard feature types, the Maplex Label Engine provides label placement options for features such as streets, contours, rivers, boundaries, and land parcels. You will use some of these capabilities to label the libraries.

1. **Ensuring that Community Anchor Institutes is the active map, on the Map tab, go to the Utah County bookmark.**

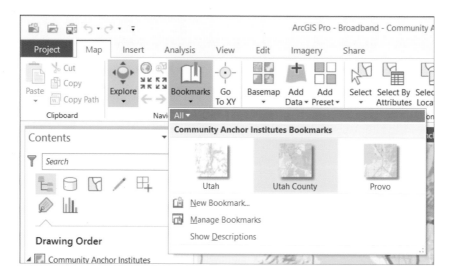

You will concentrate on labeling libraries in Utah County. Points that are densely located are more difficult to label, and you must take into consideration such factors as map scale, label font size, color, and placement.

2. **In the Contents pane, select the Libraries layer.**

3. **On the Labeling tab, in the Map group, click More to ensure Use Maplex Label Engine is checked. If it is not enabled, click Use Maplex Label Engine to enable it.**

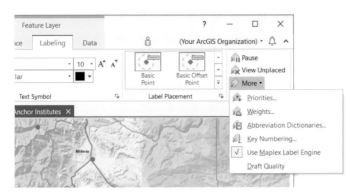

4. **On the Labeling tab, in the Layer group, click the Label button. (Alternatively, right-click the Libraries layer, and click Label.)**

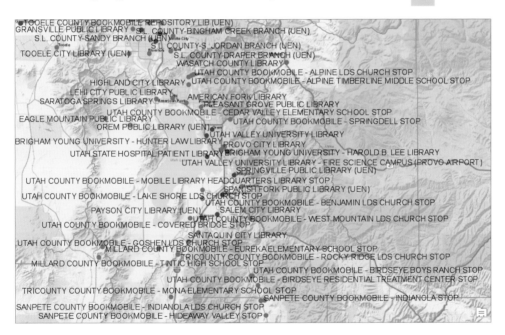

Labels are placed on the map for every feature in the layer. But labeling, in this case, is not particularly helpful because there are so many labels. The initial label placement is based on default Maplex Engine rules using the Class 1 *label class*. Label classes are used to specify detailed aspects of how labels are positioned and symbolized. Next, you'll modify the label class so that it labels only those libraries that have an unknown public Wi-Fi status.

5. **Click the label class SQL button to open the Label Class pane.**

6. **On the SQL Query tab, build a query in which PublicWifi is equal to Unknown. Click Add, and then click Apply.**

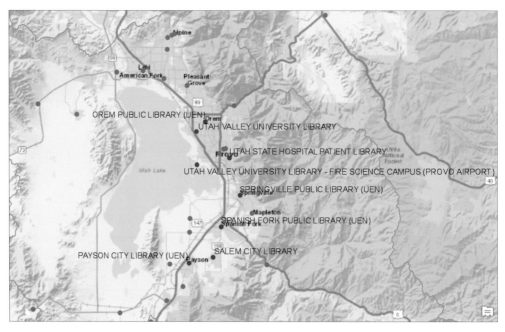

ON YOUR OWN

Optionally, use a label expression to change the labels from all uppercase letters to title case. A label expression is used to format labels using Python, VBScript, or JScript logic. Try applying the Python expression [ANCHORNAME].title() to the label class. The .title() text function formats the labels to title case (for example, SALEM CITY LIBRARY becomes Salem City Library).

The total number of labels has been reduced, and only labels for libraries that have an unknown public Wi-Fi status are shown. Next, adjust the label class symbol properties.

7. **In the Label Class pane, click Symbol.**

8. **Under the General tab, expand Appearance. Change the font to Calibri, the Font style to Bold, the size to 8 pt, the Color to Gray 60%, and click Apply.**

9. **Collapse Appearance, and expand Halo. Select White fill 50% transparency from the Halo symbol drop-down list, and click Apply.**

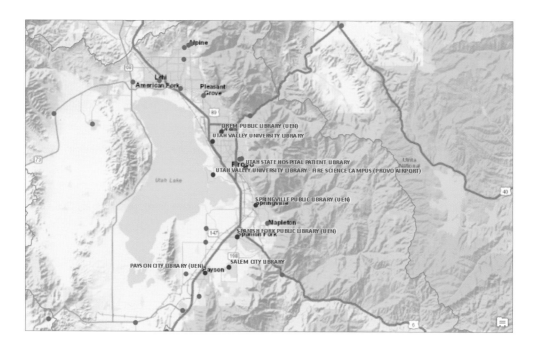

The labels now look more legible on the map. However, they still look clustered closely together. You will adjust the label class position next.

10. **In the Label Class pane, go to the Position tab, and expand Placement. Ensure that Best Position is selected in the label placement drop-down list. Change the Measure offset to Exact symbol outline.**

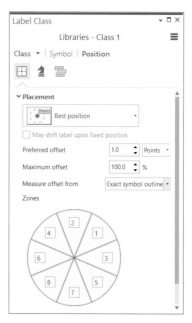

11. **Go to the Fitting Strategy tab, and expand Stack. Change Maximum number of lines to 2.** The labels now look properly placed without overcrowding. Remember that the entire map is labeled in the same way. So, if you pan or zoom to see the rest of the map, the labels will dynamically redraw to label all visible features based on this label class. To show labels at only specific map scales, a label visibility range can be specified.

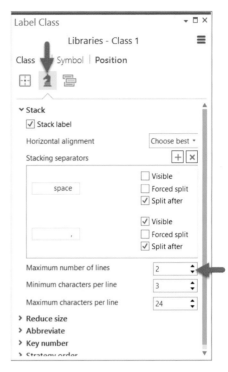

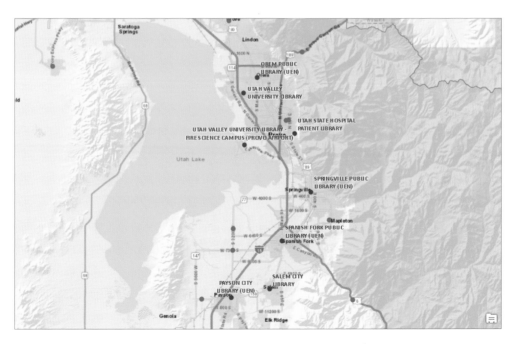

12. **On the Labeling tab, enter 1:450,000 in the Out Beyond box, and press Enter.** When the map is zoomed out beyond 1:450,000, the labels are not displayed.

You will create another label class to label more features, this time for Utah counties. This label class will provide more visual information to your map readers.

13. **Enable labeling for the Counties layer.**

14. **In the Label class pane, click Class and then the pane options menu. Rename the label class to** County Boundary.

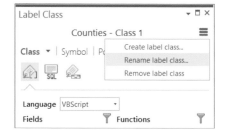

15. **Click Symbol, and under the General tab expand Appearance. Change the Font name to Century, Size to 7 pt, Color to Raw Umber, and click Apply.**

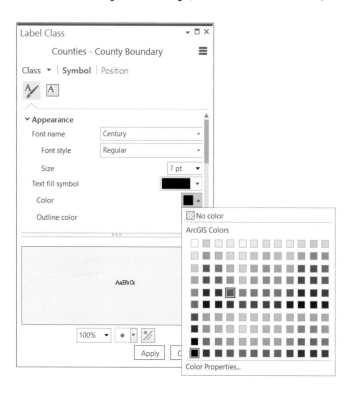

16. **Click Position, and under the Position tab, expand Placement. (Do not get confused with expanding Position under Symbol.) Select Boundary placement from the drop-down list.**

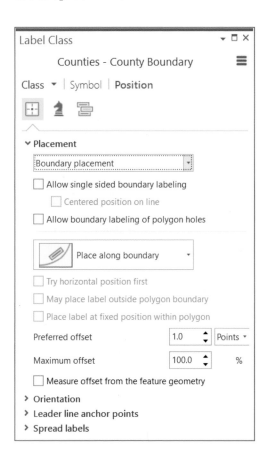

The county labels switch from the default regular placement to boundary placement, which means that they follow closely to the boundary lines. This type of labeling allows readers to easily locate boundaries on a map.

17. **Click the Conflict resolution tab, and expand Repeat. Change the Minimum interval units to Inches and choose 4.0.**

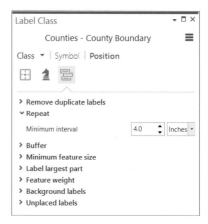

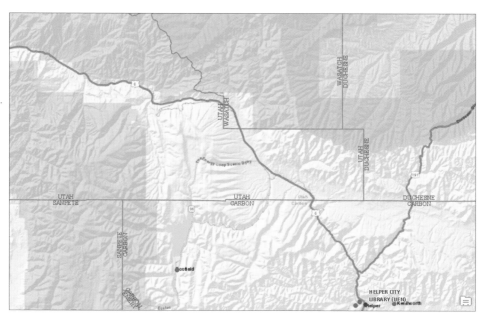

ON YOUR OWN

Create a new label class, and experiment with label placement, fitting strategies, and conflict resolution. The capability for fine-tuning labels is extensive and takes practice to get right. When you are finished, return to the County Boundary label class.

In exercise 10c, you will create a map layout that your readers can use to learn more about the broadband availability in different parts of Utah.

EXERCISE 10C

Create a page layout

Estimated time to complete: 30 minutes

A page layout includes map frames and other elements, such as scale, legend, north arrow, and more. Deciding early on the dimensions and orientation of a layout facilitates the design process. It influences your decisions on factors such as map frame sizes, map scale, text size, and how other map elements may fit.

ArcGIS Pro has rulers, guides, and a grid to help you arrange map elements on a page. You can also align, nudge (move incrementally), distribute (space evenly), order, rotate, and resize selected elements to place them where you want them.

For this exercise, you will insert map frames into the layout and add surrounding map elements that will help your readers.

Exercise workflow

- *Add a new layout to the project based on an existing size.*
- *Insert two map frames, and position them side by side.*
- *Insert a legend based on the symbology used for each map frame.*
- *Insert a scale bar for each map frame.*
- *Insert a north arrow.*
- *Add text and a title to the layout.*

Add a layout to the project

1. **On the Insert tab, in the Project group, click New Layout.**

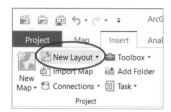

2. **From the gallery (it is in the ANSI — Landscape group), select the Letter 8.5" x 11" page size.** A landscape layout is added to the project. Like maps and data, layouts are stored in the project. You can access a layout at any time from the Project pane in the Layouts folder. Like maps, a contextual tab named Layout appears on the ribbon and contains layout-specific tools. The other tabs also contain contextual layout-specific settings.

   TIP *To change the page size or orientation after adding it, go to the Layout tab and use the controls in the Page Setup group. You can also click Properties in that group (lower-right corner), and click Size to quickly change to another standard page size and orientation; or right-click the layout name in the Contents pane to launch the Layout Properties dialog box.*

   Next, you will add content to your layout, including map frames, text, map surrounds, and graphics, using the Insert tab.

Insert map frames

Map frames are containers for maps in your page layout. They can contain any map in your project, including 3D scenes. In layout templates, the map frames can be left empty and used to point to a map later.

TIP *Changing the map extent or scale inside a map frame does not change the actual map that may already be open in your project.*

1. **On the Insert tab, in the Map Frames group, click the Map Frame down arrow, and select Community from the gallery.** A map frame that contains the Community Anchor Institutes map is added to the page.

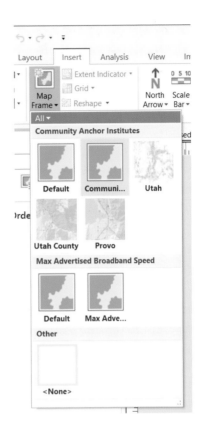

2. **With the map frame selected, go to the Format tab in the Size & Position group, and change the width to** 5 in **and height to** 6 in. **(Hint: the Size controls are on the far right.)**

3. **In the Size & Position group, click the upper-right anchor, and change the X and Y to** 10.6 in **and** 7.5 in, **respectively.**

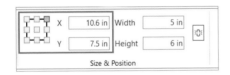

The Size & Position controls allow you to fine-tune placement on the page. Next, you will add a guide, and the Max Advertised Broadband Speed map as a map frame.

4. **Right-click the upper ruler, and click Add Guides.**

5. **In the Add Guide dialog box, type** 0.375 in **in the Position box, and click OK.** A guide is placed at the 0.375 inch position of the layout. Guides are available to help you position frames and other elements in your layout. A useful feature is to snap layout elements to guides. When you move or resize any layout element near a guide, the element will snap to the guide.

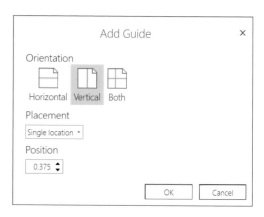

6. **Insert the Max Advertised Broadband Speed as a map frame. Use the Size & Position controls to set a map frame size width and height of 5 in and 7 in.**

7. **Move the Max Advertised Broadband Speed map frame, and snap it to the guide. Attempt to top align the frame with the other map frame.**

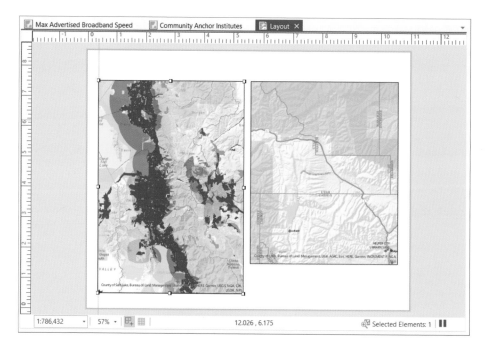

As you noticed, smart guides appear as you move the map frame to help you align it with the other map frame.

8. **With the Max Advertised Broadband Speed map frame still selected, right-click, and click Activate. Alternatively, on the Layout tab, in the Map group, click Activate.** In activated map mode, you can work with the map within the context of the page. Notice that the contextual controls change to the same commands as they do in a *map view*, and a Layout item appears in the Contents pane. Another set of navigation tools becomes available specifically for the layout.

9. **On the Map tab, go to the Utah County bookmark.**

10. **Click the Close button or the Layout link in the upper corner of the view to exit the activated map mode.**

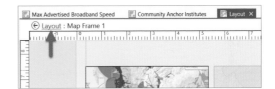

11. **Activate the Community Anchor Institutes map frame, and go to the Provo bookmark.** Both maps now show the areas of interest for your project.

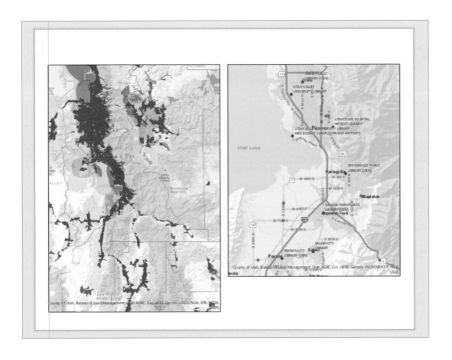

12. **In the Contents pane, rename Map Frame to** Community Anchor Institutes, **and rename Map Frame 1 to** Max Advertised Broadband Speed.

TIP *Keyboard shortcuts are useful and can help you save time when working with layouts. Generally, the layout view must be in focus to use the keyboard shortcuts. For more information, go to ArcGIS Pro Help > Layouts > Keyboard shortcuts for working on the layout.*

Insert a legend element

A *legend* will help your map readers understand the meaning of the broadband speed symbols on the map as well as the status of public Wi-Fi availability at libraries in Utah County. A basic legend consists of a patch (symbol) and a label (explanatory text).

1. **With the Max Advertised Broadband Speed map frame selected, go to the Insert tab. In the Map Surrounds group, click the Legend button.**

The cursor changes to a cross hair. You will draw a rectangle to create the legend.

2. **Click and drag a rectangle to draw the legend anywhere in the layout view. Make sure to create a reasonably sized rectangle. (A small rectangle will not show any legend elements.)** A Legend item appears in the Contents pane. Because both the Wireline_Broadband and Wireless_Broadband layers use the same symbology, it is not necessary to list the symbology for both layers in the legend.

3. **With the legend selected, go to the Contents pane. Expand Legend, and clear both the Wireline_Broadband and Counties check boxes. Alternatively, you can also remove the layers.** Next, adjust the properties of the legend.

TIP Some layout tools leave the cursor in a tool mode, such as when drawing a rectangle. To return to the select mode, go to the Layout tab, and click the Select button.

4. **In the Contents pane, expand Legend.**

5. **Right-click Wireless Broadband to open the legend's properties.**

6. **In the Format Legend pane, expand Sizing, change Patch Width to 16 pt, and change Patch Height to 10 pt.**

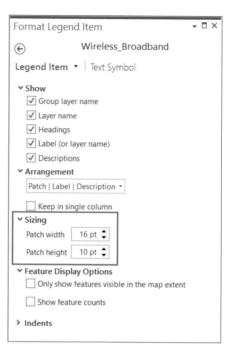

7. **Expand Show, and clear both the Layer Name and Headings check boxes.** These headings are not necessary because there is only one set of symbols in the legend. Layer and heading names are useful when multiple groups of symbols are represented.

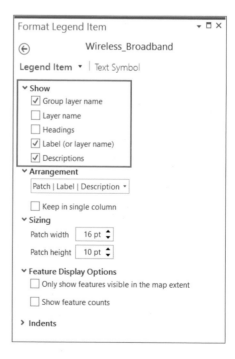

8. **Click Text Symbol to switch available options. Click Properties, expand Appearance, and change the Font name to Arial and Size to 9 pt. Click Apply.** Next, add a background for the legend so that it stands out when placed over the map frame.

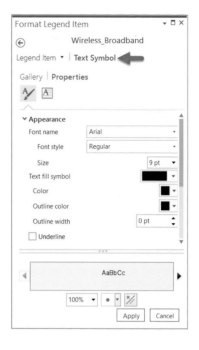

9. **Click the back button to return to the Legend options. Click the Display tab and adjust the following properties:**
   - **Expand Border, and change the border width to 0 pt.**
   - **Expand Background, click the Fill down arrow, and choose Arctic White.**
   - **Click the Fill down arrow again, and select Color Properties.**

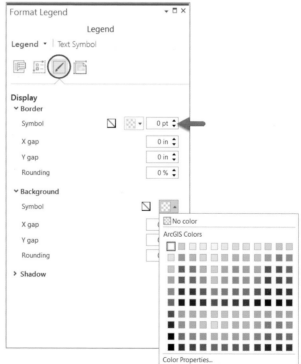

10. **In the Color Editor, change the Transparency to 30%, and click OK.**

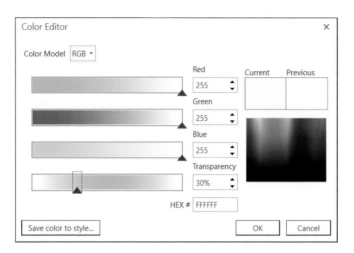

11. **Adjust the gap to improve the spacing around the patches and labels. Increase both the Background X and Y gaps to 0.1 inches.**

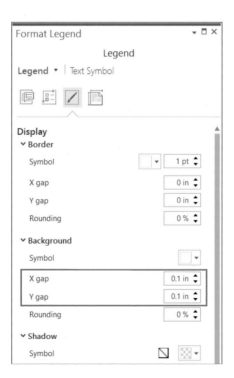

12. **Resize the legend, and position it near the upper-right corner of the map frame (an appropriate size is 1.6 by 1.5 inches).**

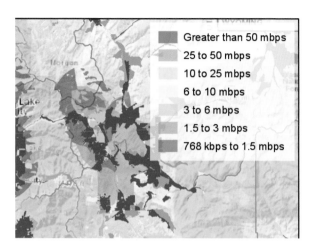

TIP *To nudge the legend into place, press and hold the Ctrl key, and use the arrow keys to move it.*

ON YOUR OWN

Follow the same steps to create a legend for the Community Anchor Institutes map. First, rename Group Heading to **Available Public WiFi***. (Hint: activate the map first, and go to the Symbology pane.) Next, adjust the group heading font, and then position the legend in the upper-right corner of the map frame.*

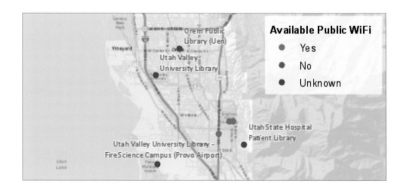

Insert a scale bar

A *scale bar* is a dynamic element that provides an indication of the size of features and distances on the map. You will add a scale bar for each map frame.

1. **With the Max Advertised Broadband Speed map frame selected, go to the Insert tab, and in the Map Surrounds group, click the Scale Bar down arrow. Select Scale Line 3 from the gallery under Imperial.** A scale bar based on the Scale Line 3 *style* is placed on the page. A Scale Bar item appears in the Contents pane. Next, modify the scale bar to suit the map frame.

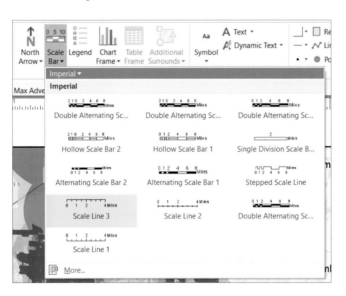

2. **With the scale bar selected, go to the Format tab, and in the Text Symbol group, change the font size to 8 pt.**

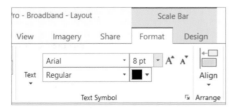

3. **On the Design tab, in the Divisions group, change Divisions to 1. In the Units group, change the label to Miles.**

4. **Double-click the scale bar to open the Format Scale Bar pane and adjust the following properties:**
    - **Similar to the legend, create a white background with 30% transparency.**
    - **Resize the scale bar so that it displays to 20 miles.**
    - **Position the scale bar near the lower-left corner of the Max Advertised Broadband Speed map frame.**

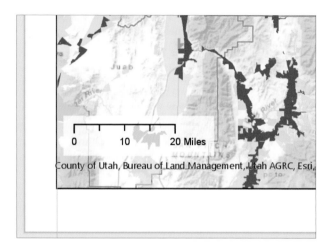

5. **In the Contents pane, right-click the existing scale bar element, and click Copy. Then right-click Layout, and click Paste.** A duplicate of the scale bar is added to the page. It sits on top of the existing scale bar. Modify it to match the map scale of the other map frame.

6. **Move the duplicated scale bar to the other map frame, and position it near the lower-left corner. On the Design tab, in the Map Frame group, select Community Anchor Institutes Map frame from the drop-down list.**

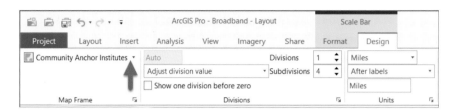

The scale bar changes to match the map scale of this map frame.

7. **Resize the scale bar so that it displays to four miles.**

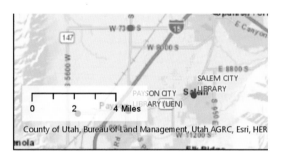

Both scale bars are configured and placed according to the map scale of each map frame. Next, you'll insert more map elements to customize your layout.

Insert a north arrow

A *north arrow* is a dynamic element that indicates the orientation of the map.

1. **On the Insert tab, click the North Arrow down arrow, and select any north arrow from the gallery.** A north arrow is placed at the center of the layout view.

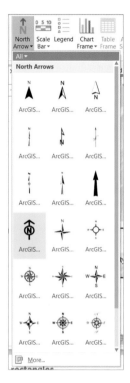

2. **Select the north arrow, and position it below the Community Anchor Institutes map frame. Resize it if necessary.**

3. **Double-click the north arrow. In the Format North Arrow pane, click the north arrow symbol. Change the fill color to Gray 60%.**

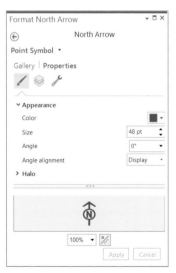

4. **Press and hold the Shift key, and select both the north arrow and the Community Anchor Institutes map frame. On the Format tab, click Align, and select Align Left. (Alternatively, if you have smart guides enabled, you can just easily align the map frame and north arrow by hand.)**

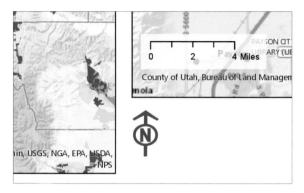

Insert dynamic text, a title, and rectangles

Now you will finish the layout with some descriptive text elements and decorative shapes.

1. **On the Insert tab, in the Text group, click the Dynamic Text down arrow, and select Spatial Reference.**

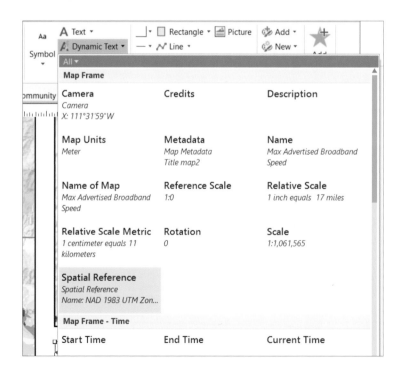

2. **Right-click the dynamic text element, and select Convert To Graphics.** Although *dynamic text* is useful, in this case there is more information than required. You will remove some of the text so that it shows only the coordinate system information.

3. **In the Format Text pane, keep only the tags so that it matches the graphic, and delete the others.** The tags contain information about current properties of the project, layout, map frame, and so on. When that property is updated, the text automatically updates. The pcs tag displays information about the projected coordinate system, the gcs tag displays the geographic coordinate system, and the projection tag displays the type of projection used.

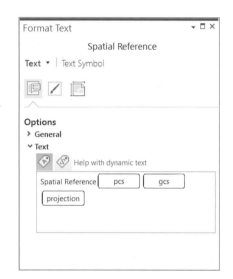

Dynamic text works through the use of tags, similar to HTML. To learn more about tags, go to ArcGIS Pro help > Layouts > Graphics and text > Modify dynamic text.

4. **Position the spatial reference information element next to the north arrow.**

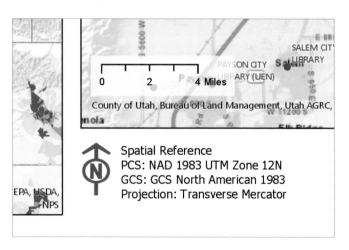

5. **On the Insert tab, click Text. Click anywhere on the layout. Change the text to** Maximum Advertised Download Speeds Utah County. **(Hint: to put "Utah County" on a separate line, press Shift + Enter.) Click anywhere to finish entering the text.**

6. **Double-click the Maximum Advertised Download Speeds Utah County text title object, and adjust the following properties:**

    - **In the Format Text pane, click Text Symbol; under the General tab, expand Appearance.**
    - **Change the font to Calibri and the font size to 18 pt.**
    - **Expand Position, change the Horizontal alignment to Center, and apply the changes.**

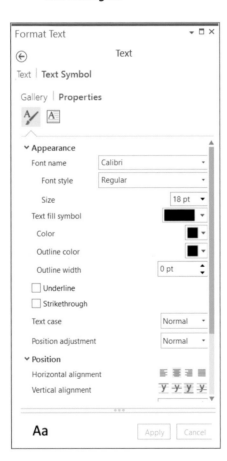

7. **Center the text element above the left map frame.**

ON YOUR OWN

Follow the same steps, and create a title for the Community Anchor Institutes map frame using the title **Public WiFi Status at Libraries Utah County.** *Optionally, insert additional dynamic text elements that may be relevant (such as user, date, or project folder path).*

8. **On the Insert tab, in the Graphics group, go to the Default Polygon Shape Symbol drop-down list, and select Gray 20% Transparent from the gallery.**

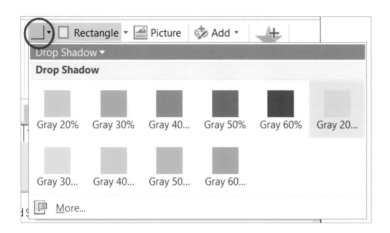

9. **Draw two rectangles the same width as the map frame (5 inches), and position a rectangle above each map frame.**

10. **In the Contents pane, move the Rectangle element below the Text element, as shown in the graphic.** Hopefully, your page layout looks good, and all the elements are where they should be. Page layouts take time to create, and when done properly, they can communicate information effectively. The next step is to share your layout.

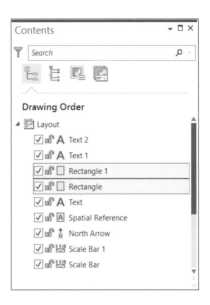

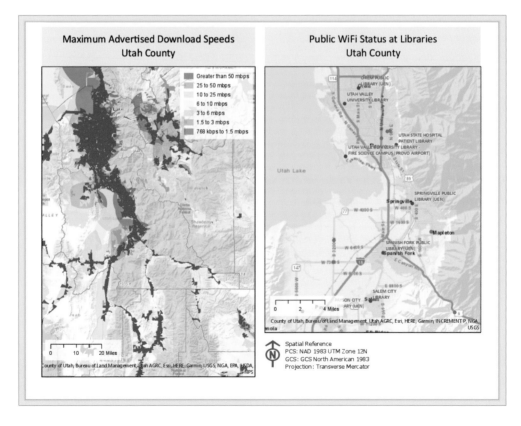

EXERCISE 10D

Share your project

Estimated time to complete: 20 minutes

After creating your layout, you can export it to a file or share it online. You can share it as a file several ways. A PDF (Portable Document Format) is one of the most popular industry-standard formats, because a PDF is editable in many graphics applications and retains map georeference information, labeling, and feature attribute data.

Exercise workflow

- *Save the map as a geospatial PDF.*
- *Save the project as a layout file.*
- *Share your project template to ArcGIS Online.*

Export a layout to a PDF

1. **Make sure that you are in the Layout pane. On the Share tab, in the Export group, click Layout.**

2. **Browse to your project folder, and name the file** Utah County Broadband Map.

3. **Click Export Options. Select PDF Layers Only from the Layers and attributes drop-down list. Make sure the "Export map georeference information" check box is checked, and then click OK.**

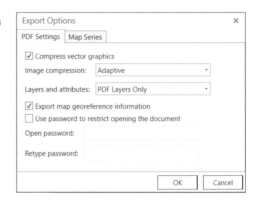

4. **Click Export.**

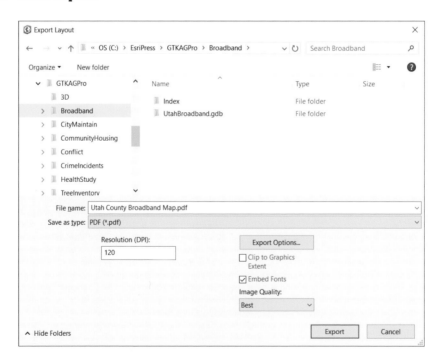

The export may take a few moments to finish.

5. **In a file browser, navigate to your project folder, and open the Utah County Broadband PDF.**

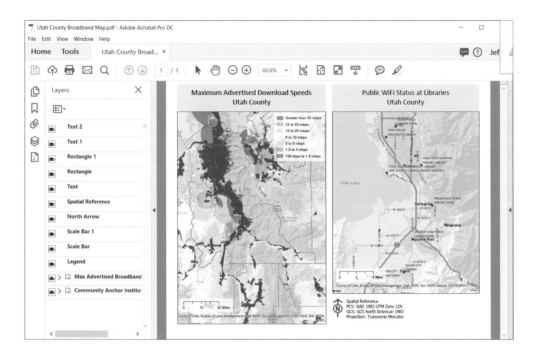

You can use Adobe Acrobat tools to find and mark location coordinates; measure distance, perimeter, and area; change coordinate system and measurement units; and copy location coordinates to the clipboard. Then you can also use these tools to show locations in several web mapping services.

OTHER EXPORT FORMATS

You can export to EMF, EPS, PDF, SVG, and SVGZ formats, which support a mixture of vector and raster data. You can also export to BMP, JPEG, PNG, TIFF, TGA, and GIF image formats, which are purely raster data. Most raster formats can include a world file that contains georeferenced information so that it can be used in ArcGIS Pro or other GIS applications. (This option is available only for map exports.) The TIFF format can retain georeferenced information internally—no external reference file is required—and is known as a GeoTIFF.

Save a layout file

A layout file includes the page, layout elements, and any maps referenced by map frames on the page. You can use a layout file to create templates or share existing layouts with others. However, data is not included in the layout file—only links to the data. So, whomever you share the data with must have the data in the same folder location, or they can simply point to the data to repair the source.

1. **Make sure that you are in the Layout pane. On the Share tab, in the Save As group, click Layout File.**

2. **Browse to your project folder, and name the file** Utah County Layout. Layout files are saved using a .pagx extension. Whomever you shared your layout file with will see the same layout as it was saved, and the surrounding map elements and dynamic text will be updated accordingly. Layout files are often small and can be shared easily, including with your organization in ArcGIS Online.

Package and share a project template

Although page layout files are useful, they do not include data. To truly share aspects of your Utah broadband map project, you can create a project template, which can be shared directly in ArcGIS Pro and uploaded to your ArcGIS Online organization in only a few steps.

Your project template will include the two maps you worked on in exercises 10 a–d, page layouts, and a connection to the database used. Optionally, you can include attachments, such as image files or supporting documentation. Depending on how you share the template, another user from your organization can open the project directly in ArcGIS Pro.

TIP *Always consider how you want project templates to be shared. Do not share sensitive data or projects with everyone in ArcGIS Online (if the "Share outside of organization" option is checked). Often, it is safer to first share to your My Content folder, and then manage which group or organization you want to share with using the Share options in ArcGIS Online.*

You can also share as a project package that includes not only all maps and layer data, but also the folder connections, toolboxes, geoprocessing history, and attachments. Project packages are an excellent way to share entire projects with colleagues, generate a snapshot of a project for archival purposes, or share the project outside of your organization.

1. **On the Share tab, in the Save As group, click Project Template.** The Create Project Template pane appears.

2. **Under Start Creating, select "Upload template to Online account." (Uploads require an internet connection. If you do not want to upload the project template, select "Save template to file" to save it to disk.)**

3. **Under Name and Location, change the name to** Utah Broadband Project. **(If you are saving it to disk, choose your project folder, and save it as** Broadband.aptx.**)**

4. **Under Item Description, type** A broadband project template for Utah counties **in the Summary box. In the Tags box, type** broadband, Utah, libraries **(comma separated so that individual tags are recognized).** Under Sharing Options, the organizations and groups you belong to are listed. (Your list may look different from the one in the graphic.) By default, My Content is checked—this is your personal workspace in ArcGIS Online, and only you will see it. For this exercise, you will share the project template to only My Content. Remember that you can change how your projects and content are shared in ArcGIS Online.

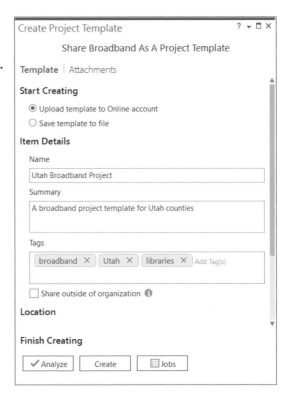

5. **Click Analyze to check for errors. If there are any errors or warnings, go back to the Template tab, and correct them.**

6. **Once validated, click Create to simultaneously create the project template and upload it to your ArcGIS Online account.** ArcGIS Pro begins to package, compress, and upload the project template. Depending on the speed of your computer and internet connection speed, it may take a few minutes.

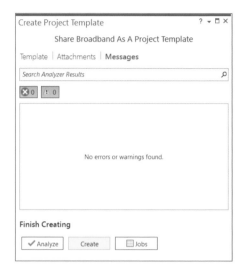

ON YOUR OWN

Using a web browser, go to ArcGIS Online and sign in. Go to your My Content workspace to see if the Utah Broadband Project successfully uploaded. From there, share with other members of your organization or any groups to which you belong.

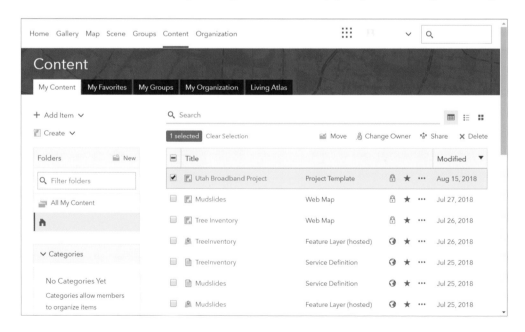

Summary

You have successfully created a layout that includes two maps of Utah County, one to show maximum advertised broadband speeds and one to show the status of available public Wi-Fi at libraries. From start to finish, you have created detailed symbology, including using a definition query to focus on specific attributes of the data and a label class to label specific libraries based on an unknown public Wi-Fi status. You also now have a foundational skill set for creating outputs that can be shared with other members of your organization. Although PDF and image exports are useful for distributing maps and information, your organization may also want to work with the data on its own using project templates. The project template you shared to ArcGIS Online is just one step of a workflow. An organization can use many additional ArcGIS Online tools to manage data and allow users to work with content. Hopefully, you are now comfortable in the ArcGIS Pro environment and are ready to begin working with your own projects.

Glossary terms

layout

label

label class

map frame

map extent

map view

legend

scale bar

style

north arrow

dynamic text

Image and data source credits

Image credits

Chapter 1

Abu Dhabi City, map courtesy of Municipality of Abu Dhabi City.

Ogallala Aquifer, map courtesy of Center for Geospatial Technology, data from US Geological Survey's Nebraska Water Science Center.

Senior Shedding in Portland, map courtesy of Portland State University; data from Oregon Health Division, US Census Bureau, Portland Metro's Regional Land Information System, US Geological Survey digital elevation models.

Chapter 3

Atlas of Heart Disease and Stroke, map created using the Interactive Atlas of Heart Disease and Stroke, a website developed by the Centers for Disease Control and Prevention, Division for Heart Disease and Stroke Prevention, http://nccd.cdc.gov/DHDSPAtlas.

Chapter 9

Continuous data map, data courtesy of the Office of Geographic Information (MassGIS), and the Commonwealth of Massachusetts Information Technology Division.

Discrete data map, data courtesy of Lake Tahoe Data Clearinghouse, US Geological Survey.

Scheid Vineyards photo, courtesy of Heidi Scheid.

Data credits

3D

Basemap courtesy of Esri, HERE, DeLorme, TomTom, Intermap, increment P Corp., GEBCO, USGS, FAO, NPS, NRCAN, GeoBase, IGN, Kadaster NL, Ordnance Survey, Esri Japan, METI, Esri China (Hong Kong), swisstopo, MapmyIndia, ©OpenStreetMap contributors, and the GIS User Community.

\GTKAGPro\3D\buildings.shp, courtesy of Department of Information Technology and Telecommunications (DoITT), New York City Department of City Planning.

Broadband

\GTKAGPro\Broadband\UtahBroadband.gdb, courtesy of the State of Utah Broadband Project, managed by Utah Automated Geographic Reference. Original data source is the National Broadband Map.

CityMaintain

\GTKAGPro\CityMaintain\Data\FireHydrants.shp, data courtesy of City of Troutdale, Oregon, Public Works Department.

\GTKAGPro\CityMaintain\Data\WaterPressureZones.shp, data provided by Oregon Metro, Regional Land Information System (RLIS).

\GTKAGPro\CityMaintain\Data\Valves.dbf, data courtesy of City of Troutdale, Oregon, Public Works Department.

\GTKAGPro\CityMaintain\Data\WaterLines.shp, data courtesy of City of Troutdale, Oregon, Public Works Department.

\GTKAGPro\CityMaintain\Data\Wells.shp, data courtesy of City of Troutdale, Oregon, Public Works Department.

CommunityHousing

\GTKAGPro\CommunityHousing.gdb\census_tracts, courtesy of US Census Bureau.

\GTKAGPro\CommunityHousing.gdb\city_boundary, courtesy of Los Angeles County Enterprise GIS.

\GTKAGPro\CommunityHousing.gdb\la_census_tracts, courtesy of US Census Bureau.

\GTKAGPro\CommunityHousing.gdb\social_services, courtesy of Los Angeles County Enterprise GIS.

\GTKAGPro\CommunityHousing.gdb\streets, courtesy of Los Angeles County Enterprise GIS.

\GTKAGPro\CommunityHousing.tbx\household_income.csv, courtesy of US Census Bureau, 2008–12 American Community Survey.

\GTKAGPro\CommunityHousing.tbx\public_housing_sites.csv, courtesy of Los Angeles County Enterprise GIS, Public Housing Sites, Housing Authority of the City of Los Angeles, Housing Authority of the County of Los Angeles.

Conflict

Service layer basemap courtesy of Esri, HERE, DeLorme, TomTom, Intermap, increment P Corp., GEBCO, USGS, FAO, NPS, NRCAN, GeoBase, IGN, Kadaster NL, Ordnance Survey, Esri Japan, METI, Esri China (Hong Kong), swisstopo, MapmyIndia, ©OpenStreetMap contributors, and the GIS User Community.

\GTKAGPro\Conflict\Conflict.gdb\ACLED_2005_2018_Africa. The Armed Conflict Location and Event Data (ACLED) project is directed by Professor Clionadh Raleigh (University of Sussex). It is operated by senior research manager Caitriona Dowd (University of Sussex). Andrew Linke is a consultant on the project (University of Colorado), while the data collection involves several research analysts, including Charles Vannice, James Moody, Daniel Wigmore-Shepherd, Andrea Carboni, and Roudabeh Kishi.

CrimeIncidents

\GTKAGPro\CrimeIncidents\CrimeIncidents.gdb\crime, data courtesy of City of Philadelphia Police Department.

HealthStudy

Service layer basemap courtesy of Esri, HERE, DeLorme, TomTom, Intermap, increment P Corp., GEBCO, USGS, FAO, NPS, NRCAN, GeoBase, IGN, Kadaster NL, Ordnance Survey, Esri Japan, METI, Esri China (Hong Kong), swisstopo, MapmyIndia, ©OpenStreetMap contributors, and the GIS User Community.

\GTKAGPro\HealthStudy\Data\IL_food_deserts.shp, census tracts: TomTom, US Census, Esri. Food desert attributes: Economic Research Service (ERS), US Department of Agriculture (USDA).

\GTKAGPro\HealthStudy\Data\IL_med_income.shp, counties data courtesy of ArcUSA, US Census, Esri. Median income attributes courtesy of US Census Bureau, American Community Survey.

\GTKAGPro\HealthStudy\Data\Obesity_Prevalence.dbf, data courtesy of Centers for Disease Control and Prevention (CDC).

\GTKAGPro\HealthStudy\Data\us_cnty_enc.shp, data courtesy of Esri, derived from TomTom, US Census.

Vineyard

Service layer basemap courtesy of Esri, HERE, DeLorme, TomTom, Intermap, increment P Corp., GEBCO, USGS, FAO, NPS, NRCAN, GeoBase, IGN, Kadaster NL, Ordnance Survey, Esri Japan, METI, Esri China (Hong Kong), swisstopo, MapmyIndia, ©OpenStreetMap contributors, and the GIS User Community.

Student-added service layer basemap courtesy of Esri, DigitalGlobe, GeoEye, Earthstar Geographics, CNES/Airbus DS, USDA, USGS, AEX, Getmapping, Aerogrid, IGN, IGP, swisstopo, and the GIS User Community.

\GTKAGPro\Vineyard\Vineyard.gdb\ned_1, NED data produced by US Geological Survey.

\GTKAGPro\Vineyard\Vineyard.gdb\ned_2, NED data produced by US Geological Survey.

\GTKAGPro\Vineyard\Vineyard.gdb\planting_sites, courtesy of Scheid Vineyards Inc., Greg Gonzales.

\GTKAGPro\Vineyard\Vineyard.gdb\property_boundary, courtesy of Scheid Vineyards Inc., Greg Gonzales.

\GTKAGPro\Vineyard\Vineyard.gdb\Soils_clip, soils data produced by the US Geological Survey.

\GTKAGPro\Vineyard\Vineyard.gdb\vineyard_blocks, courtesy of Scheid Vineyards Inc., Greg Gonzales.

World

Student-created service layer basemap courtesy of Esri, DeLorme, GEBCO, NOAA NGDC, DeLorme, HERE, Geonames.org, and other contributors.

\GTKAGPro\World\World.gdb\Cities, from Data and Maps for ArcGIS (2014), courtesy of ArcWorld.

\GTKAGPro\World\World.gdb\Countries, The World Bank: Air Pollution in World Cities (PM10 Concentrations): David R. Wheeler, Uwe Deichmann, Kiran D. Pandey, Kirk E. Hamilton, Bart Ostro, and Katie Bolt; from Data and Maps for ArcGIS (2014), courtesy of ArcWorld, US Census International Division, CIA Factbook.

\GTKAGPro\World\World.gdb\Latlong, from Data and Maps for ArcGIS (2014), courtesy of Esri.

\GTKAGPro\World\World.gdb\Ocean, from Data and Maps for ArcGIS (2014), courtesy of Esri.

Glossary

This glossary contains definitions for the ArcGIS terms used throughout the book and noted in the list at the end of each chapter. Many of these definitions can also be found here: http://support.esri.com/en/knowledgebase/GISdictionary.

ArcGIS Pro specific terminology can be found here: https://pro.arcgis.com/en/pro-app/get-started/arcgis-pro-terminology.htm.

address locator. A dataset in ArcGIS that stores the address attributes, associated indexes, and rules that define the process for translating nonspatial descriptions of places, such as street addresses, into spatial data that can be displayed as features on a map. An address locator contains a snapshot of the reference data used for geocoding and includes the parameters for standardizing addresses, searching for match locations, and creating output.

alias. An alternative name specified for fields, tables, files, or datasets that is more descriptive and user-friendly than the actual name.

altitude. The height above the horizon, measured in degrees, from which a light source illuminates a surface. Altitude is used when calculating a hillshade, or for controlling the position of a light source in a scene.

ArcGIS Pro. An application for creating and working with spatial data on your desktop. It provides tools to visualize, analyze, compile, and share your data, in both 2D and 3D environments.

aspect. The compass direction that a topographic slope faces, usually measured in degrees from north. The aspect can be generated from continuous elevation surfaces.

attribute. Nonspatial information about a geographic feature in a GIS, usually stored in a table and linked to the feature by a unique identifier. For example, attributes of a river might include its name, length, and sediment load at a gauging station. Attribute tables also store information about feature geometry, such as length and area.

attribute domain. In a geodatabase, a mechanism for enforcing data integrity. Attribute domains define what values are allowed in a field in a feature class or a nonspatial attribute table. If the features or nonspatial objects are grouped into subtypes, different attribute domains can be assigned to each of the subtypes.

attribute join. Appending the fields of one table to those of another through an attribute common to both tables. See also *spatial join* and *relate.*

attribute query. A request for records of features in a table based on their attribute values.

azimuth. In GIS software, the direction from which a light source illuminates a surface.

basemap. A map depicting background reference information, such as landforms, roads, landmarks, political boundaries, or satellite imagery, onto which other thematic information is placed.

buffer. A zone around a map feature measured in units of distance or time. A buffer is useful for proximity analysis.

cell. The smallest unit of information in raster data, usually square in shape. In a map or GIS dataset, each cell represents a portion of the earth, such as a square meter or square mile, and usually has an attribute value associated with it, such as soil type or vegetation class.

clip. A command that extracts features from one feature class that reside entirely within a boundary defined by features in another feature class.

command line. A programming environment in which the user enters commands by means of strings of text typed on a keyboard, as opposed to selecting commands from graphic prompts such as icons or dialog boxes.

continuous data. Data such as elevation or temperature that varies without discrete steps. Because computers store data discretely, continuous data is usually represented by triangulated irregular networks (TINs), rasters, or contour lines, so that any location has either a specified value or one that can be derived.

coordinate system. A reference framework consisting of a set of points, lines, or surfaces and a set of rules used to define the positions of points in space in either two or three dimensions. The Cartesian coordinate system and the geographic coordinate system used on the earth's surface are common examples of coordinate systems.

data type. The attribute of a variable, field, or column in a table that determines the kind of data it can store. Common data types include character, integer, decimal, single, double, and string.

defined interval classification. Also known as *class interval.* A set of categories for classification that divide the range of all values so that each piece of data is contained within a nonoverlapping category.

definition query. A request that examines feature or tabular attributes based on user-selected criteria and displays only those features or records that satisfy the criteria.

discrete data. Data that represents phenomena with distinct boundaries. Property lines and land-use areas are examples of discrete data.

dissolve. A geoprocessing command that removes boundaries between adjacent polygons that have the same value for a specified attribute.

dynamic text. Text placed on a layout that changes based on the current properties of an element, such as a user name, the date of a project, the file path of the project, the metadata of a map on your page, and so on.

edge. A line between two points that forms a boundary.

edit sketch. In ArcGIS, a temporary feature sketch for creating new features or editing existing feature geometry. Edit sketches that are saved become permanent.

endpoints. The points that define the length of a line segment.

equal interval classification. A data classification method that divides a set of attribute values into groups that contain an equal range of values.

extrude. To stretch a flat 2D shape vertically to create a 3D object.

feature class. In ArcGIS, a collection of geographic features that have the same geometry type (such as point, line, or polygon), the same attributes, and the same spatial reference. Feature classes can be stored in a geodatabase, shapefile, or coverage, among other data formats. Feature classes allow homogeneous features to be grouped into a single unit for data storage purposes. For example, highways, primary roads, and secondary roads can be grouped into a line feature class named "roads." In a geodatabase, feature classes can also store annotation and dimensions.

feature dataset. A collection of feature classes stored together that share the same spatial reference; that is, they share a coordinate system, and their features fall within a common geographic area. Feature classes of different geometry types can be stored in a feature dataset.

feature template. A named set of tools that creates new features. Feature templates contain properties that you can configure and define how a new feature is created.

geocoding. A GIS operation for converting street addresses into spatial data that can be displayed as features on a map, usually by referencing address information from a street segment data layer.

geodatabase. A database or file structure used primarily to store, query, and manipulate spatial data. Geodatabases store geometry, a spatial reference system, attributes, and behavioral rules for data. Various types of geographic datasets can be collected within a geodatabase, including feature classes, attribute tables, raster datasets, network datasets, topologies, and many others. Geodatabases can be stored in IBM DB2, IBM Informix, Oracle, Microsoft Access, Microsoft SQL Server, and PostgreSQL relational database management systems or in a system of files, such as a file geodatabase.

geographic coordinate system. A reference system that uses latitude and longitude to define the locations of points on the surface of a sphere or spheroid. A geographic coordinate system definition includes a datum, prime meridian, and angular unit.

geometric interval classification. A classification scheme in which the class breaks are based on class intervals that are in a geometric series. The geometric coefficient in this classifier can change once (to its inverse) to optimize class ranges.

geoprocessing. A GIS operation used to manipulate GIS data. A typical geoprocessing operation takes an input dataset, performs an operation on that dataset, and returns the result of the operation as an output dataset. Common geoprocessing operations include geographic feature overlay, feature selection and analysis, topology processing, raster processing, and data conversion. Geoprocessing helps define, manage, and analyze the information used to form decisions.

GIS. Acronym for geographic information system. An integrated collection of computer software and data used to view and manage information about geographic places, analyze spatial relationships, and model spatial processes. A GIS provides a framework for gathering and organizing spatial data and related information so that it can be displayed and analyzed.

graduated colors. A way to symbolize map features in which a range of colors indicates a progression of numeric values. For example, increases in population density might be represented by the increased saturation of a single color, or temperature differences might be shown by a sequence of colors from blue to red.

graduated symbols. A way to symbolize point or line features according to the value of the attribute they represent. For example, denser populations might be represented by larger dots or larger rivers by thicker lines.

hillshade. Shadows drawn on a map to simulate the effect of the sun's rays over the varied terrain of the land.

hot spot analysis. An analysis that attempts to identify statistically significant spatial clusters of high values (hot spots) and low values (cold spots).

intersect. A geometric integration of spatial datasets that preserves features or portions of features that fall within areas that are common to all input datasets.

intersection. The point where two lines cross. In geocoding, most often a street crossing.

join table. The table in a join operation, often a nonspatial table, that is appended to the target layer's attribute table.

kernel density. Calculates a magnitude-per-unit area from point or polyline features using a kernel function to fit a smoothly tapered surface to each point or polyline.

label. In ArcGIS, descriptive text, usually based on one or more feature attributes. Labels are placed dynamically on or near features based on user-defined rules and in response to changes in the map display. Labels cannot be individually selected and modified by the user. Label placement rules and display properties (such as font size and color) are defined for an entire layer.

label class. Label classes can be used to restrict labels to certain features or to specify different label fields, symbols, scale ranges, label priorities, and sets of label placement options for different groups of labels.

layer. In ArcGIS, a reference to a data source, such as a shapefile, coverage, geodatabase feature class, or raster, that defines how the data should be symbolized on a map. Layers can also define additional properties, such as which features from the data source are included. Layers can be stored in a project (.aprx), or saved individually as a layer file (.lyrx).

layer file. A file with a .lyrx extension that stores the path to a source dataset and other layer properties, including symbology.

layer package. A file (.lpkx) that includes both the layer properties and the source data referenced by the layer. Layer packages contain symbolization, labeling, table properties, and the data. They can be created and viewed in ArcGIS Pro, ArcMap, ArcGlobe, and ArcScene, as well as viewed in ArcGIS Online.

layout. A collection of organized elements. In a layout, common elements, including the map, map labels, a title, legend, scale bar, north arrow, captions, and additional graphics, can be arranged.

legend. The description of the types of features included in a map, usually displayed in the map layout. Legends often use graphics of symbols or examples of features from the map with a written description of what each symbol or graphic represents.

location query. Also called a *spatial query*, it is a statement or logical expression that selects geographic features based on location or spatial relationship. For example, a spatial query might find what points are contained within a polygon or set of polygons, find features within a specified distance of a feature, or find features that are next to each other.

manual interval classification. The process of sorting or arranging entities into groups or categories; on a map, the process of representing members of a group by the same symbol, usually defined in a legend.

map. The visual display of geographic data on paper or a screen.

map algebra. A language that defines a syntax for combining map themes by applying mathematical operations and analytical functions to create new map themes. In a map algebra expression, the operators are a combination of mathematical, logical, or Boolean operators (+, >, AND, OR, tan, and so on) and spatial analysis functions (slope, shortest path, spline, and so on), and the operands are spatial data and numbers.

map extent. The limit of the geographic area shown on a map, usually defined by a rectangle. In a dynamic map display, the map extent can be changed by zooming and panning.

map frame. A container for maps on your layout, which can reference any map or scene in your project. Empty map frames can be used when creating templates. The map extent of a map frame is independent of any map view in the project. A dataset can be represented in one or more map frames. In map view, only one map frame is displayed at a time; on a layout, multiple map frames can be displayed at the same time.

map projection. A method by which the curved surface of the earth is portrayed on a flat surface. Map projection generally requires a systematic mathematical transformation of the earth's graticule of lines of longitude and latitude onto a plane. Some projections can be visualized as a transparent globe with a light bulb at its center (although not all projections emanate from the globe's center) casting lines of latitude and longitude onto a sheet of paper. Generally, the paper is either flat and placed tangent to the globe (a planar or azimuthal projection) or formed into a cone or cylinder and placed over the globe (a cylindrical or conical projection). Every map projection distorts distance, area, shape, or direction, or some combination thereof.

map view. An all-purpose view for exploring, displaying, and querying geographic data. This view does not show any map elements, such as titles, north arrow, and scale bars.

mask. A means of identifying areas to be included in analysis. Such a mask is often referred to as an *analysis mask* and may be either a raster layer or feature layer.

metadata. Information that describes the content, quality, condition, origin, and other characteristics of data or other pieces of information. Metadata for spatial data may describe and document its subject matter; how, when, where, and by whom the data was collected; availability and distribution information; its projection, scale, resolution, and accuracy; and its reliability regarding some standard. Metadata consists of properties and documentation. Properties are derived from the data source (for example, the coordinate system and projection of the data), and documentation is entered by a person (for example, keywords used to describe the data).

model. In geoprocessing in ArcGIS, one process or a sequence of connected processes in ModelBuilder.

ModelBuilder. The interface used to build and edit geoprocessing models in ArcGIS.

natural breaks (Jenks) classification. A method of manual data classification that seeks to partition data into classes based on natural groups in the data distribution. Natural breaks occur in the histogram at the low points of valleys. Breaks are assigned in the order of the size of the valleys, with the largest valley assigned the first natural break.

near. An operation that calculates distance and additional proximity information between the input features and the closest feature in another layer or feature class.

NoData. In raster data, the absence of a recorded value. NoData does not equate to a zero (0) value. Although the measure of an attribute in a cell may be zero, a NoData value indicates that no measurement has been taken for that cell at all.

north arrow. A map symbol that shows the direction of north on the map, thereby showing how the map is oriented.

on-the-fly projection. ArcMap can display data stored in one projection as if it was in another projection. The new projection is used for display and query purposes only. The actual data is not altered.

open data. Data that is licensed to permit people to use the data freely. Open data can be reused, redistributed, and shared—even commercially.

operator. The symbolic representation of a process or operation performed against one or more operands in an expression, such as + (plus, or addition) and > (greater than). When evaluated, operators return a value as their result. If multiple operators appear in an expression, they are evaluated in order of their operator precedence.

organization. An ArcGIS Online organization is shared workspace composed of users, items, and groups. Users can view, contribute, and share web layers, maps, and other content.

overlay. In geoprocessing, the geometric intersection of multiple datasets to combine, erase, modify, or update features in a new output dataset.

project. In ArcGIS Pro, a collection of related geographic datasets, maps, layouts, tools, settings, and resources. A project is stored as an item in ArcGIS Online or on disk as an .aprx file.

projected coordinate system. A reference system used to locate x, y, and z positions of point, line, and area features in two or three dimensions. A projected coordinate system is defined by a geographic coordinate system, a map projection, any parameters needed by the map projection, and a linear unit of measure.

Python. A free, cross-platform, open-source programming language that is embedded in ArcGIS products and used for geoprocessing scripting.

quantile classification. A data classification method that distributes a set of values into groups that contain an equal number of values.

query expression. A type of expression that evaluates to a Boolean (true or false) value and is typically used to select those rows in a table whose values cause the expression to evaluate to true. Query expressions are generally part of a SQL statement.

raster. A spatial data model that defines space as an array of equally sized cells arranged in rows and columns and composed of single or multiple bands. Each cell contains an attribute value and location coordinates. Unlike a vector structure, which stores coordinates explicitly, raster coordinates are contained in the ordering of the matrix. Groups of cells that share the same value represent the same type of geographic feature.

reclassify. The process of taking input cell values and replacing them with new output cell values. Reclassification is often used to simplify or change the interpretation of raster data by changing a single value to a new value or grouping ranges of values into single values—for example, assigning a value of one (1) to cells that have values of 1 to 50, two (2) to cells that range from 51 to 100, and so on.

relate. Establishing a connection between two tables using an attribute that is common to both.

scale bar. A map element used to graphically represent the scale of a map. A scale bar is typically a line marked like a ruler in units proportional to the map's scale.

script. A small program or sequence of instructions, or the act of writing such a program; in ArcGIS, scripts are written using Python in a programming environment known as ArcPy.

select. To choose from a number or group of features or records; to create a separate set or subset.

shapefile. A vector data storage format for storing the location, shape, and attributes of geographic features. A shapefile is stored in a set of related files and contains one feature class.

snapping. An automatic editing operation in which points or features within a specified distance (tolerance) of other points or features are moved to match or coincide exactly with each other's coordinates.

space-time cube. A set of points summarized into a data structure by aggregating them into space-time bins that form a cube. Within each bin, the points are counted, and specified attributes are aggregated.

spatial join. A type of table join operation in which fields from one layer's attribute table are appended to another layer's attribute table based on the relative locations of the features in the two layers.

spheroid. When used to represent the earth, a three-dimensional shape obtained by rotating an ellipse about its minor axis, either with dimensions that approximate the earth or with a part that approximates the corresponding portion of the geoid.

standard deviation classification. A data classification method that finds the mean value, and then places class breaks above and below the mean at intervals of either 0.25, 0.5, or 1 standard deviation until all the data values are contained within the classes. Values that are beyond three standard deviations from the mean are aggregated into two classes—greater than three standard deviations above the mean and less than three standard deviations below the mean.

style. An organized collection of predefined colors, symbols, properties of symbols, and map elements. Styles promote standardization and consistency in mapping products.

symbology. The set of conventions, rules, or encoding systems that define how geographic features are represented with symbols on a map. A characteristic of a map feature may influence the size, color, and shape of the symbol used.

target table. The table to which new attributes are appended. Records in a target table are often referred to as *target features*.

task. A set of preconfigured steps that guide you and others through a workflow or business process. A task can be used to implement a best-practice workflow, improve the efficiency of a workflow, or create a series of interactive tutorial steps. Tasks reside in task items, which are stored within an ArcGIS Pro project.

temporal data. Data that represents a state in time. Time values are stored in a single attribute field and can be used to visualize features or phenomena at moments on the timeline.

vector. A coordinate-based data model that represents geographic features as points, lines, or polygons. Each point feature is represented as a single coordinate pair, and line and polygon features are represented as ordered lists of vertices. Attributes are associated with each vector feature, as opposed to a raster data model, which associates attributes with grid cells.

vertex (vertices). One of a set of ordered x,y coordinate pairs that defines the shape of a line or polygon feature.

web layer. In ArcGIS Online, web layers are designed for map visualization, editing, and querying and consist of two types: feature layers and tile layers. Feature layers support vector data and are most appropriate for visualizing data on top of basemaps. Tile layers are a collection of predrawn map images, or tiles, and are most appropriate for basemaps.

Task index

Sidebar topics

Chapter 2

Chapter 3

Chapter 4

Chapter 7

Chapter 10